T0320569

Cybersecurity for Everyone

Cybersecurity for Everyone

David B. Skillicorn

CRC Press is an imprint of the
Taylor & Francis Group, an **informa** business

First edition published 2021
by CRC Press
6000 Broken Sound Parkway NW, Suite 300, Boca Raton, FL 33487-2742

and by CRC Press
2 Park Square, Milton Park, Abingdon, Oxon, OX14 4RN

CRC Press is an imprint of Taylor & Francis Group, LLC

Library of Congress Cataloging-in-Publication Data

ISBN: 9780367642785 (hbk)
ISBN: 9781003124030 (ebk)

Typeset in Computer Modern font
by KnowledgeWorks Global Ltd.

Contents

Preface

Cyberspace is a critical part of all of our lives. Although we all use it, much of its infrastructure and operation is invisible to us.

The origins of this pervasive system lies in a network developed by the U.S government's Advanced Research Projects Agency (ARPA) which was originally designed to connect government and university research labs. Participants were trusted and accountable and, as a result, security received hardly any attention.

The consequences are still with us. This small network grew like a weed until it connects many billions of computational devices across the world. At every stage, design decisions were made in a way that prioritized the next great thing, and security still received hardly any attention.

So we find ourselves with an environment, cyberspace, that is, close to an essential service for many of us, but requires a significant effort to prevent the negative consequences of criminality, actions by governments for whom cyberspace acts as a cheap force multiplier, and the unthinking actions of the bored and careless.

Making cyberspace more secure is one of the challenges of our times. This is not only (or perhaps even primarily) a technical challenge. It requires actions by governments and businesses to encourage security whenever possible, and to make sure that their own actions do not undermine it.

Unfortunately, some of those in position to cause improvements in cybersecurity—lawyers, politicians—do not have the skills to understand the issues fully. This book is intended to help that situation by giving a description of the issues that is accessible to those from non-technical backgrounds.

This isn't a research monograph, so I haven't included references to the research literature. For those who want to pursue any particular topic more deeply there are plenty of web sites that will provide further information. I hope that what you read here will be enough to understand these more technical presentations.

Successful books today seem to either have titles with three verbs, on the model of "Eat, Pray, Love" or contain the word "girl". I couldn't come up with a title of the second kind, and I didn't think "Compute, Attack, Defend" really worked as a title of the first kind.

Chapter 1
Introduction

Cyberspace, the combination of computer systems, devices, and the networks that connect them, is pervasive. The most obvious part of cyberspace, and the one that most people have encountered, is the Internet, but there are many other components to cyberspace: automated production lines, separate military networks, and dark corners that are technically part of the Internet, but difficult for ordinary users to access. We as individuals use cyberspace visibly for work, for shopping, and for entertainment; but it is used less visibly, but more importantly, for keeping our societies running, in manufacturing, services, financial markets, and government.

Although cyberspace is pervasive, most of its structures are invisible to us even when we are using them. We interact with it via phones or computers, but the servers that power web sites and allow us to manage our bank accounts, to book travel and hotels, and communicate using email or social media are not easily appreciated. The "plumbing" that connects our interface devices and these services is also almost entirely hidden. Making this huge, complex system so unobtrusive and seamless from a user perspective is a major success.

However, from a security perspective, the invisible nature of cyberspace creates an issue—people have no intuition about how it works, and so cannot easily judge which actions are potentially dangerous. It's a bit like visiting a tourist-focused city and accidentally straying into a bad neighborhood simply by not realizing that such neighborhoods exist. It's hard to assess the risk of any particular action in cyberspace, and many of the "rules" come without explanation, or ar least without meaningful explanation, that would enable users to understand when rules have reasonable exceptions. For example, a common rule is "don't open email attachments from people you don't know" but this is often unworkable in practice. If you don't know that From: email addresses can be spoofed then the apparent converse "it's ok to open attachments from people that you do know" will sometimes lead to unfortunate actions.

Younger people, who have grown up with the Internet, are in some ways savvier about the dangers, but even they can easily be blind to risks and taken in by those who exploit cyberspace for their own purposes. Education about cyberspace can help, and that's the purpose of this book, but a little learning is a dangerous thing (Alexander Pope).

The problems of cybersecurity can be traced back to the early days of what became the Internet: Arpanet, a network that connected government and university computers, and which assumed that all of the participants were trustworthy and accountable. As a consequence, security issues were not considered seriously in the design of Arpanet's basic working—and many of those designs are with us still. The fundamental problem is backwards compatibility: each new development had to continue to allow existing computers, networks and systems to work together, even as the size and scope of the overall system exploded. Arpanet was also designed to survive a nuclear attack, and so was designed in a distributed way. This design lives on in cyberspace today. There is no overall management of cyberspace and decisions are made by a messy collaboration of stakeholders.

In many ways, the Internet is a victim of its own success. From that initial small set of connected computers, it grew to include more government and university computers, including internationally. In the early 1980s telephone companies got into the act, allowing dedicated connections between computers that were in different cities, and then allowing computers to connect by dialing one another and communicating data over phone lines that had been designed for voice.

The introduction of the World Wide Web in 1995 marked another shift in connectivity. Suddenly individuals had a "killer app" for getting information from businesses and governments, including those far away; and these businesses and governments needed large computers to serve these *web pages* to those who wanted them. At about the same time, email, previously limited to academics, became a pervasive way of communicating for ordinary people.

Browsers and servers developed new functionality, Servers no longer simply provided content, but enabled interaction with users, including allowing users to upload content. Once again, a mechanism designed for one use was adapted to do things for which it had not been designed, and the new uses have overwhelmed the old.

In another dimension, what changed was the *scale* of cyberspace as billions of people, more than half the world's population, acquired devices that enabled them to participate; and as servers became bigger and faster in response.

Cyberspace just grew. There was no central planning; interesting people came up with new ways to use the existing infrastructure and systems and they mostly didn't think much about the security issues they were creating in their hurry to create the new great thing. As a result, cybersecurity must work in a world where security was never designed in, and so must be retrofitted with as much effectiveness as possible. Unfortunately security isn't the kind of property that can be added in—to be really effective it must be thought about at every stage of the design and implementation. As we shall see, attempts to design countermeasures to address the security weaknesses of many parts of cyberspace are necessarily contrived and so relatively ineffective.

This is, of course, a bonanza for criminals and some national governments who are able to exploit the security weaknesses for their own purposes. There is an arms race between those who want to exploit security weaknesses and those who are trying to make cyberspace more secure. The efforts to improve security are often sabotaged by users who prefer convenience to security. A number of actors, while giving lip service to security, actually prefer considerable leakiness Many Internet businesses (and some governments) use security weaknesses to collect data about Internet users that would be more difficult to collect if cyberspace were more secure. For example, many Internet businesses try to collect phone numbers from their users, under the guise of making authentication more secure, but their real motive is that knowing someone's phone number allows access to another large trove of data about them.

One of the reasons that improvements in cybersecurity have been weak and piecemeal is that there is no central governance. Control of cyberspace is split between international bodies that manage some aspects but have no real power; the owners of the infrastructure, such as the pipes that connect countries and across countries; the (quondam) telephone companies that provide phone connectivity and also tend to own large communications infrastructure; the businesses that provide large servers (for web sites and computation) such as Apple, Amazon, and Google; and national governments. All of these stakeholders have different incentives, sometimes very different, and so developments are the result of a tug of war between competing interests. Most would claim vigorously that security is important, but this tends to translate into action only weakly if at all as other interests compete with or outweigh it. Even with the best will in the world, retrofitting security is difficult so it is not surprising that there is overall little appetite for it.

International cooperation about cyberspace is difficult. In some ways it's the first post-Westphalian[1] system. Although the Internet stretches into every country on earth, national governments have little control over it. They would like to have more, but their citizens demand access to apps and capabilities that are incompatible with national control.

A number of countries have tried to (re)construct national borders in cyberspace. The effort to control traffic that passes across such borders has had only limited success, and not for want of trying. Even if traffic could be controlled, every country still ends up using the same hardware, systems, apps[2], and ways to move traffic and so faces all of the same security issues that result. All of these pieces can become vehicles for covert activities. No country is a cyberspace island, entire of itself.

Chapter 2 provides an introduction to what is happening under the covers in cyberspace, that is, how the interfaces we use to browse the web or send email or otherwise act in cyberspace actually work. Chapter 3 discusses how content is protected by encrypting it, both in flight (as it moves from one place to another), and at rest (when it is stored somewhere). Chapter 4 describes the security issues for the nodes that make up cyberspace, the edge devices such as phones and personal computers that we all use, and also the more-powerful computers—called *servers*—that provide web pages and web-based services and the ability to carry out large computations. Chapter 5 describes the security issues for the pipes that move data around the world, from the intercontinental

[1]The Peace of Westphalia, in 1648, ended the Thirty Years War, establishing the idea that states did not interfere in the domestic affairs of other states—in that case their religious beliefs. The problem with cyberspace is that its structure crosscuts national borders and a framework for thinking about this intersectionality has not yet been established

[2]China has successfully developed its own version of platforms that are used in the rest of the world, but they have a large domestic population, and there are signs that their citizens use the international apps when they can, to evade the state surveillance built into the local apps.

undersea cables to the Bluetooth connection between your phone and your earbuds. Chapter 6 describes the security issues associated with setting up and managing the configurations of nodes and pipes so that they can work together as they should, even when something goes wrong. Chapter 7 describes the security issues associated with higher-level activities, the ones that we as users interact with directly. This includes email and the use of the World Wide Web (henceforth, the Web). Since blockchains are a hot topic, and one with substantial, complex security issues, we also discuss it.

Chapter 2
How cyberspace works

2.1 Encounters with cyberspace

Almost everyone is familiar with cyberspace in some form, but the interfaces through which we use various online systems do not necessarily reveal very much about what is behind the scenes. The three most popular interfaces are web browsing, social media, and email, and these are accessed, by most people, using phones. However, personal computers are still used for access, especially for tasks that are computational, data intensive, or require visualization.

Web browsing is the most intuitive. Web content is stored on large computer systems, called web servers. Whenever someone opens a browser and enters a URL (or clicks a link), a request is sent to the corresponding server and a web page is sent back to the browser. Although the interface creates the impression that we visit a web site, the reality is that pages from the web site visit us.

In the early days of the Web, web servers provided prebuilt web pages to browsers which were then responsible for displaying them on whatever device the user was using. The Hypertext Markup Language (HTML) describes what kind of thing each piece of the web page is (a heading, a paragraph) which enables the web browser to display the page in whatever way produced the best effect on the particular device that the user is using. For example, the browser can fit the displayed page to the available screen size,

More interesting possibilities arose when web servers were given the ability to create pages dynamically as they were requested. Now the web page that a particular user sees can be customized for that user—perhaps different for users in different countries or regions.

This ability becomes even more powerful when the user's browser can communicate individualized information to the server. The obvious use of this is search—a user inputs search terms and the search engine (a specialized kind of web server) sends back a customized page of results that match the uploaded search terms. A user can communicate with a bank web server and be presented with a page containing their own individual account balances.

The next stage happened when browsers became capable of downloading small pieces of software, along with page content. A web server, using this mechanism, can have a browser do something on its behalf. This is the key to online banking, online shopping, and all of the other interactions that are two-way, with users entering information as well as viewing content.

These mechanisms also have a negative side—they can and are used to store information in the browser (*cookies*) that can identify when the same user revisits the same web server and, increasingly, other servers as well. In other words, this rich channel of communication becomes the conduit for tracking user behavior.

In some interactions, the emphasis is more on the flow of information from the user to the site rather than the other way around; for example, uploading videos to Youtube. The mechanism is the same—the server downloads code which sets up a communication channel from browser to server that controls the upload.

Social media sites use this same basic mechanism but, instead of a (general purpose) browser they package the interactions within an app which, as much as possible, they get users to download. The advantage to the social media site or business is that the app can run *any* code, rather than the limited code that runs within a browser. The interaction can be richer because of this. This richer interaction is not necessarily a good thing—it's what makes it possible to collect data about each user (and sometimes their system) in a detailed way.

Many other businesses do the same thing: creating an app which interacts directly with their web server, rather than having users visit their web site using a browser. The app is little more than a specialized front-end to their web site, but it allows the same kind of richer interaction. This is mostly a benefit to the business rather than the user. The risk with all apps of this kind is that they may be poorly implemented and so have serious security weaknesses of their own, and they don't take advantage of the security measures that are built in to most modern browsers.

The other major way in which users encounter cyberspace is as a communication tool: emails, messaging (in multiple forms), and video chat. Some of these are symmetric interactions between one user's device and the other's, but some communication apps route all of the communications through a server. The first option is harder to implement but more private, but most tools use the second option.

It's also useful to distinguish between communications that are synchronous (the message travels immediately to the other end) such as video chat and messaging; and communications that are buffered or asynchronous, that is, can be stored at some intermediate location until the receiver is available (emails).

There are many other aspects of cyberspace that are not obvious to an ordinary user. Many businesses maintain communication with their employees, especially those who do not work from an office. They also collect data, typically in real-time, about their business activities. For example, transportation systems keep track of where every train or plane is at all times, and many other details of how they are performing and when they may need maintenance. To make possible online hotel booking sites, hotel chains must keep a constant accurate record of which rooms are available when. Retail stores keep track of exactly what has been sold at each of their stores so that they can load and dispatch a delivery truck that will arrive exactly when needed with exactly the right products on board. Electricity grids and oil and gas pipelines must track performance throughout their delivery systems and be able to control remote equipment to turn it up or down, or on or off.

Businesses also use cyberspace to communicate with one another, to provide information about supply chains or to buy from one another in the same way that we buy from online retailers. Unlike consumers shopping online, these interactions are often automated. For example, automobile assembly lines are based on just-in-time delivery of all of the components from their suppliers so that parts arrive at the assembly line just as they are needed. This reduces the costs and space required for warehousing.

Governments have a citizen-facing use of cyberspace in which they provide services to citizens online, renewing drivers' licenses, collecting taxes, and so on.

Governments also use cyberspace as an instrument of power. Almost all communication travels in cyberspace, so a great deal of the collection of intelligence has moved from radio interception into cyberspace interception. Governments also use cyberspace as a weapons system, planning (and sometimes carrying out) cyberattacks against other states.

Criminals use cyberspace, both as a way to gather information and to carry out criminal exploits. Until recently it was difficult to make money directly in cyberspace except by attacking financial institutions and financial transfers directly, but the rise of *cryptocurrencies* has made it possible to extort money from individuals and businesses, untraceably, and this has created a new set of crimes in cyberspace.

So the part of cyberspace that ordinary people see is the tip of an iceberg. There is a great deal of other communication and interaction between organizations and within cyberspace itself. The system as a whole is too complex for direct human management so there is also a layer of self-management that takes place behind the scenes to make sure that connections keep functioning even when the actions users take change.

Individuals face security issues in cyberspace, but there are many other vulnerabilities in the other, less visible, parts of cyberspace as well.

2.2 What is cyberspace?

Cyberspace is the union of all of the connected digital infrastructure in the world. The most obvious part of it is the Internet, and within it the World Wide Web. But there are also various other networks, mostly military, that are not explicitly connected to the Internet, and then there are all of the devices that are connected to the Internet but not used explicitly by humans and which we tend to forget.

Cyberspace can be considered to be made up of three different kinds of entities:

- *Nodes*, the objects that are connected to one another by networks and which carry out computations, store data, interface with humans, and interface with the physical world.

  The most obvious kinds of nodes are computers, ranging from large, powerful servers down to personal computers. But there are other kinds of nodes that are even more common. For example, most people interact with cyberspace using a phone, which is just a computer of a specialized kind. Increasingly, there are also *Internet of Things* (IoT) devices, nodes that often do not interact directly with humans, but have useful functions: video cameras, light switches, thermostats, motion detectors, and medical devices. Even less obvious are *cyber-physical systems*, such as the robots and machinery that are used in modern production lines, and the devices that control pipelines and electricity grids.

This range of devices, all containing computing capabilities and connected to networks, exist because it's almost always easier to start with a general-purpose computing device and program it to act in a certain way than to build a specialized device to act that way directly. The result is that cyberspace has almost become a monoculture since there are only a handful of computer manufacturers, operating system providers (Windows, Mac, Linux), phone systems (Android, iOS), and network switch providers (Ericcson, Cisco, Huawei).

This introduces the vulnerabilities associated with any monoculture: an attack on a particular kind of system can be successful against many targets at once. Attackers can rapidly scale their attacks and cause serious havoc. They can also try scattershot attacks on all of the nodes of a particular kind with the expectation of at least some successes.

- Pipes, the connectors that join the nodes to one another. They move data which comes in many forms: emails, requests for web pages, contents of the web pages themselves, digitally encoded voice, video, audio, software, many other kinds of data transfer, and control messages that manage the network of pipes itself.

  Pipes can be hard-wired to move data between two fixed endpoints. For example, undersea cables connect continents and can carry vast amounts of data; high-capacity networks move data around within a country, and fiber and coaxial cables move data to and from businesses, houses, and apartment buildings.

  Other pipes carry data wirelessly. These include the cell network that connects phones with the local cell tower, wifi within a few hundred square meters, Bluetooth and similar systems that connect over short distances of about two meters, and the very short range wireless connections used for RFID tags and Near Field devices over centimeters.

- Management infrastructure, which manages how nodes and pipes are organized so that data can find its way to where it is intended to go, and so that resources are shared in accordance with agreed rules. Both nodes and pipes come and go from the network as their users move around, or as systems fail or are replaced, so this infrastructure must also deal with a dynamic situation. The size of cyberspace dictates that much of this management must be automated.

People can also be thought of, in a way, as part of cyberspace for two reasons. First, much of the data moving around cyberspace is doing so because it is heading for a particular individual. So cyberspace must have ways of discovering where and how each individual is connected to the network—people must have addresses or points of presence that make sense looking out from within cyberspace.

Second, individuals act in cyberspace and we would generally prefer them to be accountable for those actions. This is a difficult trade-off: we would sometimes like to be able to act anonymously as we can in the real world, but all countries already have mechanisms to limit anonymity when it makes committing crimes easy, and similar mechanisms need to be migrated to cyberspace. The picture is muddied by the desire of large online businesses to collect lots of data about every individual to build models of them that will improve (supposedly) customized delivery of advertisements. As a result, individuals are wary of being identified when they act in cyberspace because they do not trust governments or multinational businesses not to misuse the information they gain from observing known individuals and their actions.

2.3 Nodes

Since cyberspace contains billions of nodes, there must be a way to identify which is which. This issue is more complicated than it seems because the same physical object, say a phone, can appear in multiple places over time, and sometimes at more than one place simultaneously. For example, whenever you move about (sometimes even when you go up in an elevator), your phone connects to different cell towers to maintain its connection to the cell network[3]. If someone calls you, it must be possible to work out which cell tower to use to set up the call at the exact moment that the call begins (and to keep it open if you continue to move). When you reach your office or home your phone may connect to a local wifi network. Now the phone has two points of presence in cyberspace, one

[3]Even worse your phone may be connected to two different cell phone systems at the same time, 3G for voice calls and 4G/LTE for data transfer.

where it looks like a device on the cell network and one where it looks like a device on the wifi network—at the same time.

Identifying which device is which relies on each one having a unique identifier. For phones there is a unique number, the *IMEI* (International Mobile Equipment Identity) which is tied to the hardware of the phone itself. Cellular networks can check the IMEI when a phone connects to the network and so know with absolute certainty which phone it is[4]. Countries have different law with respect to IMEIs, it is typically not illegal to change them, and the connection between a phone and its user is a weak one anyway in many countries.

Other phone identifiers are associated with its *SIM card*. SIM cards are the association between a physical phone and the provider who serves that phone. They can be swapped in and out of most phones, so they don't identify the phone as such but they get closer to being an identifier for the user of the phone, at least for users who want to receive phone calls. The first identifier connected to the SIM card is the International Mobile Subscriber Identity (IMSI) which contains details of the country, the provider, and the individual who is associated with it (if it was sold to an explicitly identified individual, which some countries require). The IMSI is used, for example, to determine whether a phone has roaming privileges on another company's network.

The second identifier associated with a phone's SIM card is the phone number. With the country and area code each number is unique across the world. This is the primary way that an incoming call is connected with the particular phone being called.

Cell networks must keep track of all of these identifiers for the phones connected to their cell towers so the incoming calls can be sent to the correct place, and correctly billed.

In parallel with the cell-phone system is the much larger Internet which connects computers (of all sizes and complexities) and also phones that are using their wifi connection capability. Each device is identified by a unique address called a MAC (Medium Access Control) address, which is six groups of pairs of hexadecimal[5] digits (such as 12:34:56:78:9a:bc). While MAC addresses are unique, they are also relatively easy to spoof, allowing one device to masquerade as another.

While a MAC address describes what each physical device is, it is convenient, and sometimes necessary, to replace one device by another

---

[4]In principle, this ability can prevent phones that have been reported stolen from being used, and of course also allows a particular phone to be tracked.

[5]That is base 16, so that the digits are written 0, 1, 2, 3, 4, 5, 6, 7, 8, 9, a, b, c, d, e, and f.

that does the same job, and to do it seamlessly. So to provide a level of abstraction above the purely physical, Internet-connected devices are all given IP addresses. In the original version, called IPv4, these are 32-bit addresses written as four groups of numbers between 0 and 255 (each based on 8 bits of the address) such as 64.233.160.1. However, because of the number of devices connected to the Internet, a new version, IPv6, was rolled out in the mid-2000s. It uses 128 bits and addresses are written using 8 groups of hexadecimal numbers.

IP addresses are hierarchical. For example, with IPv4 the first group of numbers specifies a high-level area of the Internet (usually owned by a single large business or Internet Service Provider), the next group a subregion, the next group an even smaller region, and the final group the addresses of a local set of computers. Thus knowing an IP address provides a natural, hierarchical way to find the particular device that has that address.

Finally, because these long strings of numbers are hard for humans to remember, devices can also be represented using *domain names* such as `www.google.com`. These names are also hierarchical but in a different way from IP addresses. They describe a hierarchy that is human, national, and organizational. For example, names ending in .ca are in Canada, names ending in queensu.ca are associated with Queen's University, and names ending in cs.queensu.ca are associated with the School of Computing at Queen's University. However, names that end in .com (supposed to be businesses) or .org (non-profits) do not specify which country they are, so even the hierarchical domain name structure is not completely consistent.

Because these name hierarchies cross-cut the IP address hierarchy, a non-trivial mechanism, the Domain Name Service (DNS), is needed to map between human-directed names and IP addresses.

The switches that direct traffic through the pipes are also nodes, ordinary computers optimized for the business of moving data around. They therefore have IP addresses, and all of the security issues that ordinary nodes have.

2.4 People

Although they are not, of course, part of cyberspace in a strict sense, people play an important role. In particular, as they act in cyberspace they must be identifiable in some form.

For economic transactions in cyberspace, there must be a way for transactions to be paid for, so settings like this have the tightest way of identifying which particular human is involved. This is usually done by associating a credit card with an online presence or account. Because credit card companies already verify to whom they issue cards, this ties a particular real-world person to each online account.

Increased online use of credit cards has created new mechanisms to verify that the person spending the money is entitled to. In the physical world, possession of the credit card and the PIN validates this entitlement. In cyberspace, the credit card number is easily disseminated, and the PIN can't be checked against the card directly[6]. The CVV code was added to provide an extra validation that isn't used in physical transactions and so isn't widely known.

Telephone numbers have also become personal identifiers rather than physical location identifiers. The extent to which a phone number is associated with a particular person varies from country to country depending how easy it is to buy a SIM card without providing government identification. Phone numbers obviously act as a way to find a particular person for texting/messaging.

Email addresses are also used as identifiers, not just as ways to receive information. Many web sites ask for an email address as part of their registration process, and then check it explicitly by sending an email to that account with the message containing the final steps for registering. Single-use email addresses are available for those who want to register without revealing any personal details, but some web sites check the hosts for such emails and refuse them, so there is a bit of an arms race in this area.

Finally, some online sites use handles (freely chosen names) as identifiers, rather than email addresses. Although these are chosen freely, they often contain or imply personal information, and even identity.

We all have an identity in cyberspace, often many identities and so looking out from the perspective within cyberspace, we are just another kind of edge device.

2.5 Pipes

The devices on the Internet are connected by pipes that enable data to move around. These range from the huge pipes that move data between

[6] Although portable credit card terminals are widely available in restaurants and markets—another class of Internet of Things nodes.

and across continents to the connection between a phone and a set of wireless earbuds.

As we have seen from the IP naming scheme, the Internet is organized hierarchically. To get from one part of the Internet to another requires pipes, but also switches that can take incoming traffic and pass it on in a direction that moves it closer to its destination. In other words, if the devices on the Internet are like buildings, then the structure that connects them is like the network of roads and intersections. Actually, it's more like a network of railway stations and railway yards because data can be held at switches waiting for capacity to send it on its way.

The top level of the Internet consists of wide pipes that can move a lot of data at once, and high-capacity switches that can take each incoming piece of data, decide which outgoing connection it should follow to get closer to its destination, and place it on that outgoing connection.

The higher-level switches are connected to one another, but there have also to be connections from these switches to lower-level switches, and from these lower-level switches further down, and so on to the local level. In other words, the switch for a region at any scale must connect all of the nodes in that region, but also have a connection (perhaps several) to higher-level switches.

Devices can play a dual role. For example, your phone acts as an edge device when it receives an email sent to you, but acts as a switch when it streams music from the Internet to your earbuds (and these earbuds become the edge device).

2.6 Configuration

The nodes and pipes are the bare physical structure of cyberspace, but actually using it to move and store data requires some extra structure, the rules that make it work, and mechanisms to allow nodes and pipes to join and leave.

Each path through the network must carry data between its ends. To do this there must be agreements that the nodes at the ends understand and use. These correspond to the rules of road with the edges corresponding to roads and highways.

These agreements about how traffic will move over each edge are called *protocols*. Unlike a highway, where individual drivers can make rational decisions in response to traffic conditions, traffic through pipes is just data, so all of the management must be done by the nodes at the ends. Once the traffic is launched into a pipe, it will arrive at the other end after a predictable amount of time (or perhaps not, if the pipe is cut by a backhoe).

It is unlikely that the destination node for a particular data transfer is directly connected to the node where it originates. Nodes and switches, collectively, must be able to route traffic over a sequence of edges from its starting point to its destination. Any switch must be able to receive incoming traffic, and determine which outgoing pipe it should use to move that traffic closer to its destination. To do this it maintains a *routing table* which contains entries that map destinations to appropriate outgoing edges. This is a bit like asking for directions when driving and being told "take the next right and go on for 10 miles, then ask again".

Because of the hierarchical structure of IP addresses the routing table doesn't have to have an entry for every possible destination. For remote destinations it's enough to know which way to send IP addresses that begin with, say, 173. The routing table will contain more detail about IP addresses that are physically closer, and complete addresses for the devices that are directly connected to it.

A sending node must also be aware that the receiving node at the other end of the pipe might not always have enough capacity to store and process the incoming traffic, There must be a way of agreeing about the rates at which data is sent. Sometimes this is done by an agreement in advance about how much traffic will be sent and how quickly. Sometimes it is done by having receivers acknowledge that they have received each chunk of data from senders, implicitly giving them permission to send the next chunk. Of course, this means that there is extra traffic on the edges, traffic which exists to control the operation of the network itself.

Just like real-world highways, there must be mechanisms to deal with the failure of an edge, so that traffic can be routed around the point of failure. Routing tables must therefore be constantly updated to reflect the changing landscape, and this also requires control messages.

The variability of traffic also means that entire areas can become congested, so that traffic might need to be routed around an entire region to avoid the congestion. As in the real world, this can create a dynamic situation where instability can occur—traffic is directed around a congested area creates a new congested area and leaves the original area much less congested. And then the process reverses itself. The congestion control mechanism can be subverted (as it can be in the real world) to create artificial congestion, or to divert traffic to a particular place where it could be analyzed by an adversary.

A particular node or set of nodes might want to block traffic coming from some parts of the world, or to block certain types of traffic, so the network also contains gates called *firewalls*. These can block off an entire country, for example, the so-called Great Firewall of China, but they are more commonly used to prevent only certain kinds of traffic from getting inside the subnetworks of businesses or governments. In other words, they act as filters for traffic, allowing some and discarding others.

Individuals also use a kind of gate on their phones or personal computers. These are known as *malware detection systems* or *antivirus systems* that prevent certain kinds of traffic from entering.

As well as the control functions that keep traffic flowing, and flowing in the right direction, another aspect of network management is the way in which human-directed addresses (`cnn.com`) are mapped to IP addresses, so that they can be accessed. The mechanism that does this is the *Domain Name Service* (DNS). It consists of a set of special systems that maintain mappings between one kind of name and the other.

The Internet is by far the largest network on the planet, connecting devices in all countries. However, there are other networks which are not (at least in theory) connected to the Internet. Most militaries have their own separate, more secure networks which have no direct connection to the Internet. These networks, while physically separate, use the same kinds of nodes, pipes, and control mechanisms as the Internet because it would be impractical and expensive to do otherwise. This makes them vulnerable to many of the same attacks as the ordinary Internet if an attacker can gain access to them—monoculture at work again.

The separation between networks can be illusory, so that it is sometimes possible to jump the "air gap" between them. For example, the Stuxnet malware that was used to break Iranian uranium centrifuges was introduced via USB keys that were loaded from the Internet and then plugged into computers on the separate network that controlled the centrifuges. This is a common weakness with these separate networks—there has to be a mechanism to update the software on the nodes within them, and this creates a channel for bridging to them.

Whenever a device on the Internet and a device on one of these separate networks are close to one another they may also be able to communicate using audio at frequencies to high for humans to hear, or using the camera of one to take images of the screen of the other.

2.7 Types of traffic

The key development that made the internet possible is the idea of *packet switching*. Packet switching was invented for the same reason that banks and similar organizations stopped using multiple queues in favor of one large queue with multiple servers. If data transfers happened by transferring the first entire tranche of data, and then the second, and so on, small tranches would have to wait, potentially a long time, for large ones to complete. Imagine a text message caught behind a movie download.

Instead every tranche of data to be sent over network connections is divided into quite small pieces, and each piece is sent independently. This means that small data transfers don't get stuck behind large data transfers because pieces makes progress to their destination at about the same rate no matter which tranche they came from. A small data tranche will have delivered all of its pieces sooner than a big one, but dividing them all up uses the pipe capacity fairly. (As we shall see, this technique introduces overheads in dividing up and re-assembling chunks of data which brings its own problems.)

While packet switching is egalitarian, it does mean that some kinds of data are not well served. For example, if the data transfer is a large video that is being watched in real-time at the destination (not just downloaded) it is crucial that new frames appear at a rate fast enough to be viewable. Traffic like this has a constant series of deadlines to be met, and other traffic that interferes with meeting these deadlines has a direct, observable performance impact.

Voice is often carried over networks by converting it to data (called *Voice Over IP*—VOIP) and this has the same issue. If the next piece of speech doesn't arrive in time to be played back at an appropriate speed, the audio quality sounds choppy and feels unsatisfactory. As humans, we can detect delays in the millisecond range, even if not consciously[7].

To handle these kinds of special traffic, using a mechanism that moves only packets, extra layers of protocols are used. This introduces added complexity to the operation of the pipes, which must treat urgent traffic specially, and nodes, which must make sure that they have enough free buffer capacity to keep it flowing.

[7] This is one reason why online conferencing systems feel so exhausting to use.

2.8 The Deep Web

There are parts of the Internet that are not routinely accessible to everyone. We have already mentioned networks that are not connected (at least in theory) to the Internet.

Another part of cyberspace that is less accessible to ordinary users is called the *deep web*. This contains nodes and so content that are not indexed by search engines and so are mostly invisible because they can't be found directly.

The deep web consists of two different kinds of content. The first is content that is hidden in the sense that it cannot be accessed simply by somebody deciding to visit the web page that contains it. When you log in to a web site, you gain access to content, often personalized content, that comes from a store of data at that site. For example, when you log in to your bank, you can see your account balances. Content like this cannot be seen by anybody else, and it cannot be indexed by search engines because it exists only when you, individually, are looking at it.

The second part of the deep web consists of nodes that have no public path to them. The way that search engines find all of the pages in the World Wide Web is to start from obvious places and then follow links to find new places. If a web site, say, is set up and there are no links that point to it then it cannot be discovered in the normal way. Such a web site could still be useful by giving out its address to trusted friends. (This is an example of security by obscurity since it relies on the email that contains the address not being scanned by an email provider who could then extract the address and search it.)

2.9 The Dark Web

The *dark web* is that subset of the deep web containing nodes that are explicitly trying to be hard to find, usually because they are used for something illicit and often illegal[8].

Unlike deep web sites that conceal themselves by not have explicit connections to the rest of the World Wide Web, dark web sites require special mechanisms to reach their nodes. These might be completely different ways of communicating, so that the apps used must be built from the ground up. Other, more popular, parts of the dark web look like the

[8]There is much confusion between the terms deep web and dark web, so beware.

World Wide Web, with addresses that look a bit like conventional URLs, but they have to be reached in a different way.

A dark web site, even when you know its address, cannot be directly reached by your browser. The most common part of the dark web is based on the *Tor anonymizing router*. This router accepts requests to connect to dark web sites whose addresses look like URLs except that they end in ".onion". It then uses a network of volunteer computers to randomly route the traffic through intermediate sites to break the connection between you and the web site you are connecting to. Finally Tor-based dark web sites will only accept connections from nodes on the Tor network. So between you and dark web sites is the Tor router which decouples what you do (which might be visible to someone watching your node) from the target web site you are visiting. Watching the other side of the Tor network can reveal that someone is visiting a particular dark web site, but there's no way to tell[9] who it is.

Dark web sites like these sell products such as drugs, weapons, cyber attacks, stolen credit cards and credit card numbers, and are used for human trafficking. They are also used for more benign purposes such as allowing protest movements to communicate and plan, and there are even dark web versions of conventional web sites for those who don't want their online activities tracked.

There is another part of the dark web (the darker web) that uses its own protocols to communicate. This of course requires specialized software at both ends (a dark web browser and a dark-web web server) to make sense of what is being communicated.

2.10 The World Wide Web

For many, the World Wide Web *is* the Internet. This is because web browsers can do many of the things that ordinary people in cyberspace do, and do so within a single environment. This idea has been made explicit in the Chromebook, a laptop that has only a web browser interface to the Internet. We return to the Web in Chapter 7.

2.11 Social aspects

Although cyberspace is a technical artifact, it exists within and, in many ways, enables our social interactions and this necessarily has an effect on the way it functions and is used.

[9]Well, actually...

The original basis for the Internet was Arpanet, developed by the U.S. government to facilitate research and connecting a small set of universities and research labs. Much of the design of the Arpanet survives into the Internet even though this is wildly inappropriate, because the Internet developed incrementally. There have been several iterations of research projects to design a new Internet from the ground up, but they have all run up against the sunk costs of existing systems.

In particular, the Arpanet connected trusted sites so very little attention was paid to making it secure. Network security has had to be entirely retrofitted.

The allocation of IP addresses (and domain names) is managed by international consortia, so you can't just connect a computer to the Internet. For ordinary citizens, their IP addresses are managed by their Internet Service Providers who might either allocate a fixed IP address to each user (really, their connection point to the ISP's internal network) or generate one dynamically each time a computer is connected. (This is further complicated because a home modem or router might present a single IP address to the outside world, but generate local IP addresses for each computer and phone in a household.)

However, this allocation process, while it has some structure, does not provide a simple mapping between IP address and location in the world, For example, a large organization might be allocated a large range of IP addresses, say 100.200.*.* and might allocate these to their computers in many different places. And, of course, if ISPs allocate IP addresses dynamically, there is no necessary relationship between an IP address and a street address.

So networks are quite chaotically structured, and there is very little of what might be called "management" except in the most basic sense. If something is wrong there is nobody to complain to. Progress is made by collaboration between supranational bodies, national governments, and large players such as big ISPs and multinational businesses. There is no cyberspace police force to enforce even the simplest of secure behavior.

The greatest social force, however, that introduced insecurity to the Internet is that convenience almost always trumps security. This continues to be a powerful force. In setting after setting, when individuals, businesses, and governments must make trade-offs between these two properties, convenience wins. Thus while the Internet began with little security, subsequent developments did not introduce security where they could have because of considerations of performance, cost, lack of imagination, and the need to produce results against deadlines. As cyberspace scales and becomes more pervasive, most of the cybersecurity issues that we now see will become more impactful, and no doubt there are new issues that we are not aware of yet.

Another issue that causes security problems in cyberspace is that so much of what goes on is both hidden from view and unintuitive. This is partly the success of the Internet; people can use a web browser to access the world's information without really understanding anything about what happens behind the scenes to give them that experience. Contrast this with a transportation system such as trains, stations and tracks. We might not understand much about signalling systems or how a locomotive works but at least we have some intuition about how trains are formed from carriages, how trains must wait for a green light, and how long it might take to get from one place to another. Without this intuition, it is difficult for ordinary people to make informed decisions about how to act.

Most of cyberspace is self-managing in the sense that ordinary data traffic moves from place to place without human intervention. Indeed much of what goes on is too complex for direct management. However, there must be places where humans control aspects of cyberspace, albeit at a high level of abstraction, and ordinary users of cyberspace also "control" it in the sense of deciding to go here or go there, do this or do that.

This interface between humans and cyberspace creates a new vulnerability that is known by the umbrella term of *social engineering*. Social engineering refers to all of the ways in which cyberspace can be manipulated by manipulating the actions of humans as they interact with it. This can range from the way in which social media businesses create traffic using stickiness and clickbait[10]; to the ways in which malware can be inserted into a computer by getting an email recipient to click on a link; to obtaining a password by claiming to have forgotten it when talking to a help desk.

2.12 Governance

As we have seen, the Internet has a kind of loose governance structure that is slightly American-centric because of its origins. Much of the infrastructure is owned and managed by large Internet Service Providers

[10]Whenever a media site has a headline in the form of a question, you can reliably assume that the answer is "no" and you don't need to read the story.

(ISPs) or telecommunications companies. The necessity to remain consistent with all of the other parts of the Internet constrains any given infrastructure owner. One of the biggest issues making any kind of improvement to cyberspace is the sheer number of stakeholders, with a wide variety of understanding of the issues.

Net neutrality is the principle that organizations that carry data, that is, those who provide the pipes, do not provide preferential service to particular other organizations. This principle has been under fire recently as, for example, streaming entertainment services would like stronger guarantees of how much of their data can be moved and how fast.

The global and slightly anarchic structure of the Internet does not fit well with Westphalian international legal principles. This has resulted in attempts by governments to carve the Internet into national subinternets that they can control.

The best known of these is the so-called Great Firewall of China. The Chinese government controls traffic into and out of China (and within China as well) using a combination of rules embodied in filtering devices and a very large group of human supervisors. However, even this well-resourced and draconian approach leaks because of the need to provide tourists, and especially international business people, with access to their home nodes while they are in China. As a result, many international hotels in China are connected to the Internet in the ordinary way. Leaks like this have enabled various workarounds to evade the Great Firewall.

Other countries, notably Russia, have experimented with cutting their national section of the Internet off from the rest of the world. This would require turning on the same kind of filtering used by the Chinese at short notice, or perhaps even cutting connections completely. The difficulty with ideas like this is that it is hard to make sure that all the doors have been closed without actually trying it. Short tests have been carried out, but the economic impact makes it difficult to mount a thorough test.

The fundamental problem is that the structure of the Internet cuts across national boundaries, so that a web site used in one country need not necessarily be in the country, or even on the same continent.

Another area where Internet structure and national boundaries cause an issue is the question of where data is stored. The rules for the protection of data and data privacy vary from country to country so whether data can be accessed by, say, law enforcement or intelligence organizations will depend on where it is stored. Concerns about privacy means that, increasingly, countries are legislating where data about its citizens can be held.

2.13 Security issues

As we discussed in Chapter 1, most of the security issues in cyberspace arise because security was almost never considered in the design of the systems on which it is built. However, even if security had been part of the design, there are a number of difficult issues in the design of a complex, distributed, incremental, world-spanning system.

The typical starting point for thinking about security in cyberspace comes from the design of medieval castles. Castles are defined by walls, intended to keep attackers out, and often arranged in rings so that penetrating one ring did not mean that the whole castle has been conquered. The problem with using this as a model for security is that castles necessarily have gates. Historically more castles by far were conquered via their gates, often opened by someone on the inside, than by battering down the walls. The same is true in cyberspace: security mechanisms often resemble building thicker and higher walls, without paying enough attention to the fact that data has to move in and out of regions and systems, and the places where this happens are often poorly defended, Of course, the walls are more obvious and more impressive than the gates, in cyberspace as they are in the real world.

Security wasn't considered in most parts of cyberspace when it was designed and built. Unfortunately, retrofitting security is extremely difficult. A major part of the difficulty comes because of the need for backward compatibility—all of the systems must continue to operate even when some part of it is upgraded to be more secure, and there's no overarching authority that can mandate installing the new improved part everywhere. There are many parts of cyberspace that are using computers and software whose flaws are widely known but which have not been updated or replaced because of cost, ignorance, or lethargy. Sometimes this heterogeneous nature, mixing new and old security regimes, introduces new vulnerabilities, so that security often seems like two steps forward, one (or two) steps back.

Some of the major players are taking responsibility for forcing updates to systems that increase their security. For example, Microsoft makes it much harder not to install updates to Windows automatically; Google and Mozilla both update their browsers behind the scenes; phone manufacturers force system updates; and Amazon updates Kindles, all without (much) user action.

Forcing updates seems like an inherently good idea, but it is actually a two-edged sword. Updates to address security issues often break other parts of running systems, and this can be disastrous for businesses that rely on constant availability. As a result, organizations tend to try and block automatic updating; and so security vulnerabilities for which a fix is available are often still present in important systems, providing targets of opportunity.

A more fundamental problem is the asymmetry between attack and defence. Any part of cyberspace must successfully defend against *all* possible attacks. An attacker only has to find one attack that works. This forces defenders to spend resources on defence, most of which is motivated by hypotheticals and so a hard sell to those providing the resources. Attackers may also have higher levels of motivation. The security staff of an organization is working for a salary; an attacker often has a more-personal or direct purpose.

Another significant problem is the pull on cyberspace stakeholders to indulge in *security theater*. This is when an organization imposes security processes that appear to be increasing security but actually do little or nothing to actually improve it. Often this means pushing processes that address low-risk but visible vulnerabilities while ignoring less visible but more substantial risks. A common example is a requirement to change passwords regularly, despite considerable evidence that this causes more problems then it solves (people change passwords back to what they were by cycling through the required number of possibilities, or write passwords down).

2.14 Non-benign use of cyberspace

Cybersecurity aims to prevent exploitation of cyberspace for purposes other than normal use. This range from vandalism by the bored, to money-making criminality, to use by states as levers of international policy. Some uses are not technically criminal but are social destructive.

For criminals, security vulnerabilities create new ways for them to commit crimes, primarily crimes that make money (rather than violence). Some of these are online version of traditional crimes, for example, stealing money from people's bank accounts rather than mugging them. The difference between online and offline crimes of this kind is scale—it takes the same effort to steal $100 as $100,000—and timing—transfers can happen in a few milliseconds.

Online crimes also tend to be more invisible. The victim of a mugging knows that it has happened, but the victim of a pickpocket does not, at least for a while. Online crimes are more like pickpocketing, but we are not used to keeping track of online resources in the same way that we keep track of wallets. Because online information and knowledge are digital, stealing them means taking a copy, so there need be no obvious sign that anything has happened.

The reach of email and other forms of messaging also makes it possible to find the gullible with much less overhead than in the real-world. In fact, online criminal scams often make themselves seem even less credible than they are inherently because this ensures that only the truly gullible get taken in. Various kinds of confidence tricks (Nigerian princes) have flourished in cyberspace, and they continue to make money for their perpetrators.

Cyberspace has also created new kinds of crimes that rely on its new capabilities. A good example is extortion based on ransomware. A ransomware attack encrypts the files on a system and demands payment from the system's owner to decrypt the files again.

The existence of cyberspace has also made a difference to criminals because of the communication with one another than it enables. Criminals can find other criminals with particular skills with whom to collaborate. Criminals can also learn new skills online. Products that are useful in committing crimes can be purchased online much more easily than in the physical world. Online shopping can provide tools such as lock picks that are illegal to possess and so hard to buy and sell in the real world. Criminals can also buy software tools ("cyberattacks in a box") that they can use in cyberspace to carry out crimes.

Cyberspace has also become an arena for national power. In a passive form, it is a platform for intelligence gathering, since almost all data is carried within cyberspace, rather than by radio. All governments use this to spy on other countries, and many also use it to spy on their own citizens when they are in other countries (and often at home as well). Cyberspace can also be weaponized, so that one country can attack another country by disabling the target's online activities, which could have a huge impact given how dependent we are on online activities; or by using cyberspace to carry out attacks on physical systems (which are almost always controlled using computers).

States also carry out low-level disruptions of other states via these other states' own citizens, using disinformation to destabilize them by attitude change or by trying to sway elections.

There are two uses of cyberspace by businesses that are at least problematic. First, social media businesses that rely on grabbing attention (for example, to serve ads) have developed ways to make their interfaces more enticing so that users interact with them more often and for longer. This has had the unfortunate side-effect of causing psychological harm to people both from the negative content that is exposed to keep them interested, and because the time and attention that is grabbed take away from other aspects of life.

Second, cyberspace makes it possible to collect large, detailed descriptions of everyone's activities. These are widely distributed to businesses that use them to model individuals. Often, this modelling is ostensibly aimed at serving them better ads, but the data never goes away and so can, and inevitably will, be used for less savory purposes.

Some of these negative activities in cyberspace are corruptions of normal processes, but some of them require misusing cyberspace. For criminal actions, legal remedies can be used but they suffer from the mismatch between the national nature of legal codes and the transnational nature of cyberspace. There is little need for criminals to operate in the country into whose jurisdiction they fall. This is compounded by the difficulty in keeping legal codes relevant in the fast-moving world of cyberspace. Law makers do not, generally speaking, understand cyberspace and so law making is too slow and too unfocused to help as much as it should.

In the international arena, the same problems are writ large. International law does not apply to most of the problems of cyberspace, and states are free to do as they please, compounded by the difficulty of attribution when a state does act.

Chapter 3

Encryption and hashing

In the early days of the telegraph, telegrams were charged by the word so businesses developed codebooks that could express something which would take many words to say ("the Captain is insane") with a single code word ("enbet", a real example). Of course, this had the added advantage that the meaning of communications was obscured but that was not their primary purpose. Nevertheless, because telegraph operators see the content of everything they transmit, there were circumstances where the use of a code was relevant. For reasons of security, the code phrase to indicate the death of Queen Elizabeth II is allegedly "London Bridge is down".

When telephone systems replaced the telegraph, there was less need to worry about the content of communications being intercepted (at least once telephone operators ceased to play a role). Interception was still possible, but it required physical actions and, in many countries, a legal process.

In cyberspace, communications of all kinds flow across pipes and through nodes in an unpredictable way, and often the equipment through which it passes is in a jurisdiction that is not that of the endpoints. It quickly became pressing to be able to conceal the content of data transfers from interception and prying eyes. Enter encryption.

Encryption maps data from the form in which it originally exists (called *plaintext*) into a new form (called *ciphertext*) using a mapping with these two properties:

- Without knowing the key, it is hard to reverse, in the sense that it would take many years of computation to recreate the original content; and
- There is no relationship between the structure of the original data and its encrypted form. For example, two almost identical plaintexts should be mapped to be very different encrypted forms.

The essential property of encryption is that knowing the encrypted form (the ciphertext) provides, in a practical sense, no way to recreate the original data (the plaintext) unless you hold the key.

Encryption makes it possible to move data around in a network, even if the network isn't trusted, because being able to see the encrypted data doesn't let anyone except the sender and receiver discover the content of what is being communicated.

3.1 Private key encryption

There are two main approaches to encryption: *private key encryption* and *public key encryption*. In private key encryption, there are two keys, one which is used to encrypt the plaintext data, and the other which is used to decrypt to ciphertext, getting back the original plaintext. Usually these two keys are the same, or at least closely related to one another so that both ends can use the same key to transmit to one another. For example, the Caesar cipher uses an encoding in which each letter of the alphabet is replaced by the one, say, 5 to its right (so "a" becomes "f") during encryption. During decryption, each character of the encrypted message is replaced by the one 5 to its left. As long as both ends agree that the offset is 5, they can communicate with one another using this simple encryption method.

Encryption systems have been in use for millennia, but they often relied on the secret of the method, rather than the secret of the key. After all, breaking a Caesar cipher only requires trying every possible offset, which was practical right from its first use. Any substitution cipher that works by replacing each single character by another in a regular way is easy to attack by looking at character frequencies. For example, the most common letter in English is "e", so whichever character is most common in the encrypted message is probably the encrypted version of "e".

Encryption such as that used by the German Enigma system in World War II used character substitution but the substitution changed with every character of the message. This makes it much, much harder to decrypt but not, as it turned out, beyond the computational capacity of the new computing devices developed at Bletchley Park in the 1940s.

The lesson learned from this was that it's not enough for an encryption system to look as if decryption without the key is difficult; it must be provably difficult.

Both ends of the communication must know their (identical or closely related) keys and must keep them a secret. If someone else learns either one they can get access to encrypted data (and can also generate data that will appear to have been encrypted by one of the ends and so look like it must have been sent by them).

This introduces a thorny problem—how to get the common key(s) to both of the endpoints—called the *key distribution problem*. The keys can't just be transmitted as data because the whole point of using encryption in the first place is that unencrypted data can be intercepted—and keys are an especially valuable kind of data. They could be sent encrypted in some other way, but this just pushes the problem back another step, since the keys for that encryption method must have been distributed somehow. Key distribution really requires some kind of trusted side-channel, but these are necessarily hard to construct in settings where there is already a need for encryption. Private-key systems have fallen out of use because of the key distribution problem, except for settings where there is time to distribute the keys physically.

3.2 Public key encryption

The second main approach to encryption is more common because it has a number of attractive properties compared to private-key systems. It is called *public key encryption*. In this scenario, the key used for encryption and the key used for decryption are different, so different that one cannot be inferred (computationally) from the other. However, they work in the same way as before: if a plaintext message is encrypted with the encryption key then the ciphertext can be decrypted back to plaintext using the decryption key.

A node (or person) A has a private key, which must be kept secret, and which is used for decryption. A also has a public key which does not need to be secret and which can be used by anyone to encrypt messages to A. Letting this key be widely known isn't a problem because the only way to use it is to encrypt messages that only A can decrypt; and A's private key can't be worked out from knowing its public key.

So anyone who wants to send a message to A can look up its public key, use it to encrypt the message, and send it to A, who is the only one who can decrypt it. The two nodes do not have to share any common keys. A can create both keys, keep the private key and make the public key public[11].

3.3 Digital signing and digital hashing

We already know that decryption is an inverse to encryption (starting with a message, encrypting it, and then decrypting it produces the original message). Often the structure of encryption systems means that encryption is also an inverse to decryption. In other words, if a message is first transformed by applying the decryption mechanism, then applying the encryption mechanism will once again produce the original message.

When this is true, public key techniques can be used for *digital signing*. If A creates a legal document and then uses its (private) *decryption key* to encrypt it, then the resulting document is tied to A in a strong sense. Anyone can use A's public key to *decrypt* it; if the result is a sensible document then it must have been encrypted by A (the only one who knows the right key). If the result is random garbage then the document was not encrypted by A. This acts like A's signature: only A could have made it happen, it's tied to A and A can't repudiate it, and nobody can change it (say, to try and make it someone else's signature) without breaking it.

Digital signing can be used to deal with one of the problems of public key encryption systems—guaranteeing that A's public key really does belong to A. If anyone can manage to distribute their own public key but make it seem as if it belongs to A then they can read all of the messages intended for A. The solution uses *certificate authorities*, organizations who collect fees for verifying and validating public keys.

A sends its public key to a certificate authority, together with a claim of identity ("I really am A"). The certificate authority uses one of a number of mechanisms to check that A really is who it claims to be, and the issues a document contains A's public key and A's identity, digitally signed using the certificate authority's private key. Anyone who wants to check A's public key can use the certificate authority's public key to unwrap this document and make sure the identity and public key match.

[11] There is one remaining issue: how can someone else know that what purports to be A's key really came from A but we ignore that.

Of course, this means trusting that the certificate authority's public key really does belong to it; and this can be done by relying on a higher-level certificate authority (and so on). So this process is about increasing confidence—corrupting a certificate authority is harder than corrupting *A*—but the process comes with no ultimate guarantees.

The cost of getting a certificate is one major reason why small organizations tend not to use public-key encryption when they could. But certificates are an ongoing mess, because they tend to expire at awkward moments, and most people, when they encounter a certificate problem, continue to do what they were going to do anyway.

Another technique that uses ideas from encryption is *digital hashing*. Here the goal is to map some data into an encrypted form, but the encrypted data is forced to be of a restricted size. We are not interested in being able to decrypt the data so we only need a one-way mechanism, that is, we use an encryption key but there is no matching decryption key. The essence of the idea is that the encrypted form, of the data, the digital hash, is a compact representation of the entire data because of the property of encryption that any small change in the plaintext should produce a large change in the ciphertext.

Because there can be arbitrarily many sets of initial data, but the number of encrypted forms is limited to a fixed size result, this mapping isn't guaranteed to produce a unique hash for all initial sets but collisions are extremely unlikely if the fixed size is not too small. For example, if the fixed size is 32 bytes then there are 2^{256} possible digital hashes, which is far more than the number of documents that will ever exist.

What we require is that the mapping mixes up the relationship between inputs and outputs in the same way encryption does. Two similar initial pieces of data must produce very different hashes, and it must not be possible, knowing some plaintext and its hash, to infer another plaintext that hashes to the same value.

A digital hash is a kind of digest of the entire content of a set of data, but one that can't be manipulated without knowledge of the key. It is used to create a compact representation of a document or the code of an app so that its integrity can be checked. An app developer can provide both the app and its digital hash; someone who downloads it can recalculate the hash from the app code, and check that the answer matches the original hash to verify that the app hasn't been altered maliciously or corrupted in transit.

3.4 Encryption in use

The basic purpose of encryption is to allow private communication between two parties, that it to eliminate the possibility of eavesdropping. It is needed in cyberspace because there is so little control of where and how data will travel on its path from source to destination.

Encryption is designed to make it impossible in practice to decrypt data without knowing the appropriate key. This has not always been true even in systems where there are (supposed) proofs of the difficulty of decryption without knowing the key. There have been examples of government signals intelligence lobbying for encryption schemes that have known but subtle flaws so that brute force attacks can be used without knowledge of the key and decrypt in reasonable amounts of time.

The implementations of encryption systems that are theoretically expensive to decrypt can introduce shortcuts unintentionally, making them vulnerable in practice. And people can be careless with their keys, leaving them on their systems where they can be stolen.

There is regular discussion by government and law enforcement going like this: "Encryption provides an opportunity for malefactors to conceal their communications with one another. Therefore, encryption systems should be provided with backdoors that allow law enforcement to decrypt such communications (with some judicial oversight)". These backdoors are basically the decryption keys.

You can see now that it isn't possible to have a backdoor and at the same time maintain the strength of the front door. The only way to decrypt messages in a well-designed encryption system is to use the decryption key. For law enforcement to create their backdoor they would have to insist that they hold a copy of everyone's decryption keys. These keys would become one of the highest value targets on the planet since getting hold of them would make huge amounts of data available in their unencrypted forms. The vault in which they were stored would have to be protected accordingly, not just from outside attackers, but from insiders who might be suborned, or find it irresistible to decrypt the emails or health test results of their favorite celebrities. The amount of trust in government technical abilities to protect these keys, and their unwillingness to use them except as legally enabled is far beyond that of any country's citizens.

So content can be protected cryptographically for everybody, or for nobody, but it's not possible to protect only good communication while allowing bad communication to be read by law enforcement or intelligence. There is no middle ground.

This is also another place where the global reach of the Internet clashes with the wishes of national governments to hold copies of keys. There's almost nothing to prevent individuals, let alone multinational businesses, from getting their encryption keys in another jurisdiction; and it's hard for a government to demonstrate that they have done so. Proposals to hold decryption keys in some kind of escrow founder for these two practical but unavoidable reasons. Unlike most other aspects of our lives, our cyberspace existence is much less constrained by national borders. This will be an ongoing theme.

Unfortunately, encryption can conceal the *content* of communication, but it cannot conceal the *fact* of that communication because it has to be possible for untrustworthy intermediaries to deliver each piece of data. This creates another kind of vulnerability, not as critical but potentially important.

At the very beginning of the 20th Century, the British Empire developed the *All Red Line*, a telegraph network that passed only through territory that the British controlled precisely to prevent interception. The equivalent cannot be done in cyberspace because the endpoints of a communication cannot control the path that it takes. Thus an adversary can, in principle, keep track of how much data is flowing between two endpoints, and how often, even if the content itself is encrypted. This is called *traffic analysis* and can be very revealing, especially if the endpoints do not take it into account[12].

As we shall see, encryption plays an important role in defending cyberspace but there are still plenty of ways of subverting it.

[12]As a real-world example, foreign intelligence organizations watch the number of pizzas delivered to the Pentagon after hours as a way of judging whether there is a defense crisis.

Chapter 4
Node security

As we have seen, the nodes in cyberspace consist of both computers and, in a sense, of people. It's useful to divide up the computers in cyberspace into several categories.

First, are the computers (and phones) that we use to interact with online information, and to communicate with one another. We might call these *edge nodes*.

Second, there are the computers, usually much bigger and more powerful, that are used by organizations to run their businesses. These include back-end processing such as keeping track of your money, running automated factories, planning how and when to resupply retail stores, and carrying out data analytics to understand complex systems. Some of these computers are implicitly visible to ordinary users, for example, the web servers that provide web pages to your browser, or travel reservation sites, but many are less visible.

Third, there are *clouds*, large groups of computers working together and available to businesses to use instead of owning and managing their own computational resources. The important different between clouds and other computational resources is that clouds are designed to be *shared* which creates new security issues.

Fourth, there are devices that contain Internet-connected computers but which do not have a directly usable interferference, *Internet of Things* devices. These are mostly sensors or actuators, for example, smart thermostats, surveillance cameras, smart lightbulbs, or earbuds.

Some work almost completely autonomously but most are controlled remotely from somewhere else on the Internet. For example, electrical transformer switches are controlled remotely from a control room; a smart lightbulb is controlled from an app on someone's phone.

Although it's natural to consider nodes as *computing* resources, there are also nodes whose primary purpose is to store data and provide it when asked. These may be specialized (for example, a database server) or special forms of data clouds.

There are estimated to be well more than 20 billion Internet-connected devices, so there are far more nodes than people.

4.1 Getting access to nodes

One of the first issues of security for nodes is who is allowed to access them. We all have experience of this with edge nodes—our personal computers and phones are protected so that only we can access them using mechanisms such as passwords or biometrics. One of the biggest security weaknesses in cybersecurity is that the mechanism for getting legitimate access to a node—*authentication*—is weak, and often, for Internet of Things (IoT) devices, extremely weak.

We can divide authentication mechanisms into three kinds: something you know, something you have, or something you are.

Something you know

The most common example of something you know is a *password*. You know it, the system you're using it to authenticate with knows it (in some form), but nobody else is supposed to know it.

The system that you're authenticating with should not store your password in its plaintext version, because then anyone who managed to breach the system would gain access to all of its accounts. To avoid this, passwords are stored in digitally hashed form. When you enter your password, the text you entered is immediately hashed using the same mapping and the result compared with the stored version. If they match, you are authenticated.

A breach of the system might allow all of the hashed versions to be taken, but because of the properties of digital hashes there is no straightforward way to map these back to the passwords that generated them. Sadly, there is a way, although a more expensive way, to work out what the passwords were. There are only a few standard algorithms to do the digital hashing necessary to create the hashed passwords, so an attacker can start by applying one of these algorithms to "a", then to "b", and so on using longer and longer words. Eventually, a word will appear that maps to the known hash and this must be the password.

That's why the advice for passwords is to make them long (because many possible text strings must be tried before the right one is found) and contain unusual characters (because this increases the variety of text strings that must be checked). It doesn't take much sophistication by an attacker to try common password choices first, and then words from a dictionary, and then dates, before resorting to a brute force search. That's why the common advice is to create passwords that are not in these predictable categories. Note that an attacker can precompute the hashes of all of the potential words shorter than a certain length, so the incremental cost to break into a new account is very small.

There is therefore an arms race between the increasing computational power of attackers and the consequent requirement for increasing complexity of passwords. At present, increasing computational power is helping attackers, while the demands of remembering ever more obscure passwords works against ordinary users. This is compounded because many users have passwords to several sites, perhaps even hundreds, and it's implausible to remember them all. Unsurprisingly, users resort to using the same passwords for many sites, to using short passwords, or to writing them down in places where others have access to them. Using passwords as an authentication mechanism is an approach that is close to failure because of the mismatch between the need for complex passwords and the cognitive demands it puts on users, even with assistants such as password managers.

Some social media businesses are trying to become authentication intermediaries. Many sites allow you to log in using your social media credentials. However, this is far from disinterested. For these social media sites this becomes one more way in which they can track you as you visit other sites on the Internet.

Many businesses try to make authentication by something you know stronger by including questions to which only you (in theory) know the answer. At first, these were questions such as your mother's maiden name, but in an online world this is often easy to discover. A wider range of questions is typically used now but users have to decide whether to use the correct answers, which makes them more guessable, or use fictitious answers, which even they may not remember.

Secret sharing is a development of something you know. Here both you and the system to which you authenticate know *different* pieces of information, which must be put together to create the "something you know" that allows authentication. As a trivial example, the system could generate a random value between −90 and +90, your secret is a letter of the alphabet, and the correct response is to name a city at the latitude given by the random value that contains your letter of the alphabet in its name.

Something you have

This approach requires a user to have and/or present some object in order to authenticate. There is a considerable range of such devices, with varying levels of sophistication.

Our physical keys are a simple example. Having the key is the only authentication you need to be able to unlock and access a real-world door or container.

You can think of a credit card which you can use to pay by tapping as another simple form of authentication by something you have. You authenticate yourself as someone allowed to run up a bill on the card simply because you have it (which is why you can't run up a *big* bill)[13].

Another large set of examples are the keyless devices that are commonly used for vehicles and doors. When these are close enough to a reader, the reader allows access to the relevant system. A reader might be attached to a computer system, perhaps in a USB slot, and the system will allow access as long as the device is close enough. Since phones are ubiquitous in much of the world, phones themselves can play the role of keyless access devices for other computers, or for physical access.

More sophisticated devices, often called security dongles, can be plugged into a USB slot and can be interrogated by different kinds of software to authenticate not just use of the computer but even of particular pieces of software on that computer. The dongle acts as a physical licence to allow access to the software. For example, it may contain a licence code for the software and a unique identification number so that accesses can be tracked.

Because the pathway between dongle and software can be attacked, or the dongle itself can be replicated, in software or hardware,

[13]To run up a bigger bill you must combine something you have (the card) and something you know (the PIN). This is called *two-factor authentication*.

dongles often contain a secure processor and encrypt the communication between software and dongle, even though this communication is only happening within a single system.

Authentication using something you have becomes more problematic when the authentication is to a remote system. This requires setting up a secure path from the reader to the remote system, which can be done, but care is required.

The other risk for using something you have as an authentication mechanism is that you can lose it or leave it at home, and so be locked out of whatever it authenticates; and it can be stolen and used by someone to masquerade as you.

These devices can also be given or lent to someone else who then has all of the access that goes with the device. This is sometimes convenient, but it also carries security risks.

Something you are

This approach uses properties of your body, *biometrics*, as the authentication mechanism, for example, your fingerprints, handprints, earprints, irises, face, or voice. At first glance this seems attractive, Because these biometrics are always with you, you can't forget them (something you know) or leave them at home (something you have).

A biometric has to be converted into something that can be stored in a computer system and, like passwords, a new presentation of the biometric has to be mapped in the same way and the result compared to the previously stored result. This stored encoding of the biometric has some of the same problems of stored password hashes—they become a target—with the added problem that the mapping isn't typically a digital hash and so may be reversible. Biometric encodings have to be designed so that they are invariant to small changes: holding a finger in a slightly different position, developing calluses, or growing a beard, say. This means that they are doing a rough match, and so this prevents true digital hashing.

Biometric authentication systems have two underappreciated but serious drawbacks. The first drawback is that they are often relatively easy to spoof. Fingerprints began to be used to identify criminals at the beginning of the 20th Century, and the ease with which they could be faked was first pointed out within a few years in the detective novel, *The Red Thumb Mark* (1909). In some systems, even today, putting plastic film over a fingerprint reader, and the pressing down on it causes it to register the most recent real fingerprint. Irises can be spoofed from a high-quality picture of someone's eye provided that it is placed behind a hole in a white sheet of paper. Face recognition is not very accurate to begin with, but can be spoofed by a mask derived from a photo. Voice recognition is also not very accurate, and changes for the same individual when they have a cold or allergies.

Biometric systems also fail because not everyone has fingers, hands, ears, eyes, a face, or a voice, and so there must always be a backup authentication mechanism. This increases costs because of the need to provide a second mechanism at every place where biometric authentication is used.

The second drawback is that we typically have *one* of each of these biometrics. If the (usually encoded version) of a biometric is compromised, there is no way to issue a new one.

When a biometric authentication has to be done remotely, to allow access to a web site say, then the encoding of the biometric has to be done at the point where it is collected, and then the encoded version transmitted to the web site for matching, This creates a vulnerability at the collection point, since the encoded version can be intercepted, and the replayed later or at a different access point. This can be worked around, as it is for passwords, but devices that collect biometrics are not always mature and so can be vulnerable to misuse.

Something you know brings mild reassurance that it is the authorized person who is authenticating. Something you are brings much greater, but not complete reassurance. The weakness of something you have is that there's no necessary link between its use and the person who is authorized to use it.

These mechanisms can be combined (called *two-factor authentication*) to try and get the best of their properties and avoid the worst. However, the more authentication channels the more problematic it becomes for human users, and the more likely that they will become annoyed and find a workaround.

4.2 Malware

If only properly authenticated users were able to use their computational devices and software, then many cyberattacks would be much more difficult. Unfortunately, people often do the cyberspace equivalent of opening gates to those with malicious intent, allowing them authenticated access to their environments.

Allowing malicious software (*malware*) to be placed on a computer is, of course, not something that ordinary users do deliberately, but there are a number of pathways by which it can happen. Usually, it is the result of an unwitting action by a user. It is common to blame users when this happens, and there is no doubt that education can make a big difference; but the design of much of cyberspace makes it far too easy for attackers to nudge ordinary users into doing something unsafe.

One of the common pathways is that the user downloads some software or an app that contains concealed malware, hidden inside its overt functionality. For apps downloaded onto phones, app stores try to make sure that the apps available do not contain malware, but their filtering process does not always work. For personal computers, it is easy to download software from any web site and so there is a correspondingly greater risk. Operating system developers are trying to move to a model that provides more vetting of downloaded software but this is not yet common, and manufacturers are using this extra security as an excuse to collect more data about users to resell to social media businesses.

Another common way to introduce malware is to attach it to an email, say, masquerading as data. For example, an email attachment may appear to be a document of some kind but actually contains executable code that is, or installs, the malware. When such malware is attached to emails sent at random to large numbers of email addresses then it is known as *phishing*. Emails may also contain links to web sites that download malware as a consequence of clicking on them.

Increasingly malware-bearing emails are targeted to particular recipients, when it is known as *spearphishing*. Considerable effort may be expended in making the emails seem plausible: faking the sender's address to be one that the recipient might expect, and making sure that the apparent content of the attachment is real so that when it is opened the recipient might not even realize that it has installed malware. Spearphishing can also use social engineering, making it seem that the email comes from the recipient's boss, or even the company CEO.

The increasing prevalence of phishing emails is having a serious impact on communication since, at any level of sophistication, it is still possible to be fooled. We can see the beginnings of this with people reaching out using another modality, such as a text message, to ask if an email attachment is legitimate.

Some forms of malware, called *worms*, can spread from system to system using the underlying mechanisms that control cyberspace traffic. When this happens a system can acquire malware without a user doing anything. As a result, worms can spread very quickly.

There is also a sense in which malware can target people rather than nodes—cyberbullying, scams instigated using emails, and the spread of misinformation.

4.3 What does malware do?

Most computer systems divide their operations into three different kinds: those that can be done by any user, those that can only be done by the operating system (and so which require administrator accounts with special permissions), and those that are done when the system is first turned on. The actions that can be done by any user are the most restricted; users can use the software on the computer but typically cannot install some kinds of new software, and cannot see or change critical files. The operating system can do anything in the whole running system. Because of this, operating system accounts (administrator accounts) are restricted to systems staff who are trained, including in security. In phone systems, the operating system capabilities are not available to users at all, remaining under the control of the operating system developer (unless the phone is rooted). Finally, the actions that are done when the system is turned on are especially powerful because they happen before any checking can be done.

Malware can be targeted at any of these three layers. User space malware is the easiest to install, because actions of an ordinary user can create the opportunity to install it surreptitiously, but as a consequence it has the smallest opportunities. This is the reason why users are often encouraged to always carry out their ordinary work using an account like this, even when they have full control of their system and could use an administrator account for everything.

Malware targeted at the operating system level is harder to install, because it requires a higher level of permission, but provides greater opportunities. Malware targeted at the code that executes when a device is turned on is the hardest to install but is, of course, the most powerful since it controls the way in which the device is configured.

The ease with which malware can be removed also reflects how difficult it was to install in the first place. Malware detection software can usually find user space malware and often malware in the operating system, but finding and removing malware in the code that runs when the device is turned on is much more difficult.

It is also possible that malware is installed in the hardware itself. This kind of malware requires installation as part of the manufacturing process, making it much more expensive, but also much, much harder to detect. This is typically the realm of governments.

Malware can be used for a range of purposes, from vandalism all the way to non-kinetic military attacks. Some of the goals of those who distribute malware are:

- Making money. The most obvious motivation for deploying malware is its use by criminals to make money. This turns out to be harder than it looks, but criminals are inventive and they have found ways. The most successful so far has been *ransomware*. Here malware encrypts the files of the target machine using a key known only to the attacker. The malware then informs the user of how much money to pay in order to get the file system unencrypted again.

  In a system where there are proper backups, ransomware is little more than a nuisance but there are a surprising number of systems where this is not the case. Unsurprisingly there have been a steady number of ransomware attacks against small to medium sized businesses, and also against larger targets such as Britain's National Health Service.

  As in most cases of extortion, it is not a good idea to pay ransoms in response to attacks. The cyber-machinery to carry out a ransomware attack can be purchased by low-level criminals as a black box and it likely that most criminals who carry out ransomware attacks are unable to actually decrypt the file system even if they wanted to.

- Intellectual property theft. Many businesses have information that is valuable to their competitors or would-be competitors. This can range from the details of new devices or plans, to the amount they plan to bid on a contract. There are incentives for a competitor to use malware to discover such data and send it outside of the target organization where the competitor can use it (called *exfiltration*). Unlike ransomware, this kind of malware tries to remain as unobtrusive as possible so that it can carry out the exfiltration without discovery. It may even delete itself when its job is done so that it leaves as small a trace as possible.

- Protest. Cyberspace provides an alternative to the real world to protest the actions of others, especially organizations. One popular way to do this is to deface web pages by inserting malware into web servers, since these are the most visible online parts of many organizations. Malware can also be used to change passwords, shutting legitimate users out of their accounts, or to delete random files. These have little long-term impact but are annoying because system staff have to remediate them. Some forms of protest are against national governments, and their appropriateness depends on your view of whether the government is legitimate or not.

- Denial of service. A more serious technique that is often motivated either by protest or economic pressure is a denial of service attack. Here the idea is to cripple a cyberspace service by deluging it with so much bogus traffic that legitimate users cannot use its services. In the real world this would correspond to sending a crowd to a store in such a way that actual customers cannot get inside. Such an attack is expensive in the real world but much easier in cyberspace where it is cheap to create bogus traffic.

  A direct denial of service attack consists of generating enough traffic to a particular destination that it, or its surrounding network environment, fail because of overloading. It is difficult to generate enough traffic directly (and such an attack could be traced back to its origin) so it is more usual to insert malware on a number of other computers. The sole purpose of this collection of computers is to generate traffic in high volume and from many different directions. The malware on these computers is designed to stay hidden and only act when their role as generators of traffic is required; such systems are called *bots*. Internet of Things systems are particularly popular targets for turning into bots since they are not usually well protected, they have enough capability to generate traffic, and nobody is watching them closely enough to detect a change in behavior.

  A *distributed denial of service* attack is triggered by getting all of these bots to begin simultaneously to demand something from the target system. Often criminals collect bots over time and rent them as a service to anyone who wants to use them for a denial of service attack. Creating and managing a network of bots (a *botnet*) is another way to monetize malware.

Targets for distributed denial of service attacks often respond by presenting an address to the outside world that can be dynamically mapped to different servers behind the scenes. Early distributed denial of service attacks were successful, but the ability to switch which actual server is responding to the next request for service makes it increasingly hard to generate enough traffic to overwhelm a target.

- Direct financial theft. Some malware is designed to collect credentials when a user, for example, logs into their banking web site. Criminals can then use these credentials to transfer money directly to accounts that they control. This represents one of the easiest ways to monetize malware directly, but banks tend to be sophisticated about fraud and are likely to block suspicious transfers quickly. This kind of theft also leaves traces that can lead back to the criminals since they must be able to access the accounts to which the stolen money was transferred.

  Stolen credentials can also be used for online shopping, using a different delivery address from which the criminals can pick up the purchased products, but this is riskier because it takes time for purchases to be delivered, and the deception may be discovered in the meantime. Of course, criminals do not use their own addresses but addresses of convenience or empty houses, tracking deliveries so that they know when packages arrive. (This is one of the reasons why online shopping sites send you an email whenever you buy something—to make sure it was you.)

  The rise of cryptocurrencies such as Bitcoin has provided a way for criminals to get money untraceably, but this requires ordinary people to convert their financial-system money into cryptocurrency and send it to criminals.

- Exerting national power. Because of cyberspace's involvement in the economic and intellectual life of the world, it is natural that states regard it as an avenue to exert power and press their strategic national aims. At the same time, they can observe other states doing the same thing and infer their strategic aims as well. The two main ways that states leverage cyberspace are:
  - Collecting intelligence. Because almost all communication travels through cyberspace, observing it is the major way in

which states carry out *signals intelligence*. Data carried by separate (military) networks is harder to collect but may still be possible because of bleeding across the supposedly secure barrier between them, and also by tapping directly into these special networks, Doing so under the ocean is preferred because it can be done in international waters, and because it is much harder to notice that it is taking place.

The signals intelligence problem in cyberspace is the converse of what it was historically—rather than having to look for *any* interesting data, so much data can be collected that it's difficult to separate the wheat from the chaff. Malware implanted in the computers, phones, or at least neighborhoods of high-value targets lessens this problem since most of the data collected there will presumably be relevant.

Intelligence can also be collected from data stored in other countries using the exfiltration techniques that are used for intellectual property theft. In some countries, government agencies carry out data exfiltration both for national purposes and to support their local industries.

- *Active cyber operations*. Governments also increasingly use, or plan to use, malware as a weapon. Part of the attraction is that, should such malware be discovered, attributing it to the state that placed it there can be difficult (at least with enough certainty to respond). Malware like this can also be placed in position but not enabled until it is needed, so it acts like a trojan horse.

  There have been a number of uses of malware in support of strategic state aims, from causing oil pipelines to explode and catch fire to the Stuxnet malware which destroyed Iranian uranium centrifuges.

There are also a number of ways in which nodes are used to affect the humans associated with them (rather than affecting the nodes themselves).

- *Cyberbullying*. Software platforms often convey information that is intended to disturb recipients. This can be either directed to specific individuals, a common problem for youth but not limited only to them; or directed to another group at large where it is known as hate speech or abusive language. Motivations are often obscure but self-loathing or othering have been advanced as possible explanation.

- Scams mediated by messaging. There are many crimes that use cyberspace as their means of reaching large numbers of vulnerable or gullible people. This can range from spam emails that promise various unlikely things such as enlarging body parts or inheriting fortunes, to scams where the sender pretends to be a government department and tries to collect money in the form of owed tax or fines. A variation on this theme is malware that causes malfunctions on the node, and then offers to sell the user a solution that will get rid of the malfunctions.

- *Disinformation.* Software platforms are also used to convey information intended to affect a political outcome, either generally to sow doubt about the democratic process generally, or specifically in favor of or against particular political candidates.

There is also a secondary market for criminals in supplying cyber tools and exploits to other criminals, often via dark-web marketplaces.

So far, we have discussed malware in the context of individual computers, but malware can also be used against servers and clouds. These larger systems are harder to attack in the sense that they are managed by trained people who are less likely to introduce malware accidentally. On the other hand, clouds at least are usually shared among many separate users, and malware in the space of one user can, in certain circumstances, interact with the space of another.

Ransomware is beginning to be used as a attack against these larger systems. Malware designed for intellectual property theft can also be installed on servers and clouds to look for interesting content to exfiltrate.

Internet of Things device have also been attacked. One major reason is to turn them into bots that can then be used to mount denial of service attacks against other systems. But IoT devices can also be used for extortion. For example, video baby monitors can be attacked so that they make horrifying or annoying sounds in a baby's room, smart lightbulbs can be made to flash randomly or refuse to turn on. As more and more IoT devices are placed in settings where their behavior matters, such attacks will become more common. For example, house doors can be unlocked remotely and heating and cooling systems turned off (or on) by criminals remotely. Even being able to tell that a light or television is off could provide information to enable a physical-world burglary.

Most malware will leave a hook in a system that it has infected so that subsequent intrusions are easier. This is even more typical for large systems because they are much juicier targets. These hooks can be hidden in such a way that, if the malware itself is discovered and removed, the hook may still lie undiscovered.

4.4 Direct attacks

The attacks we have considered so far have been either (a) on Internet-facing systems that can be attacked externally, for example, by denial of service, or (b) attacked because of actions by an authenticated user of the system who has opened a gate to let the attack in, for example, because of phishing. But what can an attacker when the target system has limited access and/or doesn't have normal users active on it?

For example, a web server is Internet-facing, but there are not usually ordinary users logged into it. The server that manages a database of business-critical information is not Internet-facing, and also usually doesn't have ordinary users logged into it—it is accessed remotely, but only from within the business.

So how can systems that are deep within the subnetwork of an organization be attacked? They have to be attacked using vulnerabilities in the pipes that connect to them and vulnerabilities in the nodes themselves.

Nodes are open to attack because they are extremely complex, and there are inevitable mistakes made in design and implementation that create vulnerabilities. Nodes are regularly updated to provide new functionality, but many of these updates are to address security vulnerabilities that have been discovered. If nodes are regularly updated then these vulnerabilities are removed, and an attacker can only use a *zero day attack*, one that relies on a vulnerability that has not yet been discovered and fixed by the system owner. For this reason, zero day vulnerabilities are valuable commodities for criminals and governments, all the more so as they reveal their existence when anyone uses them, and so tend to have a short shelf life.

Pipes cannot be easily and regularly updated because they are a shared construct that many different nodes use. A pipe update is a bit like us all agreeing that we will speak Esperanto—it requires a lot of agreement and is very costly. So pipe vulnerabilities are important enablers of attacks on nodes.

Attacking a node deep within an organization's network is a complex problem because it may require attacking intermediate nodes in order to build a bridgehead to reach the target system. We will return to this after the discussion of pipes and configuration.

Chapter 5
Pipe security

The pipes that convey data from one place to another look, at first glance, as if they should be simple—data in at one end, and data out at the other. The reality is more complex.

Consider a train track between two towns. At its simplest, it's just the continuous set of rails running from one to the other. But getting people and freight from one town to the other requires more than just the rails—there must be rolling stock to carry different kinds of content: passenger cars for people, container cars, liquid-holding cars specialized for chemicals or milk, bulk carrier cars, track inspection vehicles, and many more types. Some of these can move themselves (for example, railcars) but other must be pulled by locomotives. Trains are made by assembling these pieces into ensembles that travel down the tracks as a unit, and so there must be space to collect them at one end and redistribute them at the other. There must be a way to detect that a train did not arrive at its destination (because of a derailment or engine failure). Sometimes trains run over tracks that belong to different organizations, or even different countries, without necessarily knowing it.

The situation is the same for pipes in cyberspace. Pipes carry a mixture of different kinds of traffic with different requirements and, because of packet switching, these different kinds are interleaved. The pipes have no intelligence, so all of the decisions about what happens in exceptional circumstances are the responsibility of both ends, which must therefore be able to agree.

The sets of rules that define what traffic looks like and how it flows are called *protocols*. They constitute the rules of engagement for the nodes that each pipe connects.

Similar rules of engagement exist when we speak to one another. At the lowest level, speech is just a sequence of vibrations. At the level above this, we categorize short time-slices of vibrations as phonemes. At the level above this, we categorize groups of phonemes as words and at the level above this, we can talk about the language we are speaking, and ideas such as speeches or conversations. A telephone can carry speech but it is only aware of speech as vibrations. A voice recognition system, such as many automated attendants, must understand speech at the higher level of phonemes and words. An intelligent agent or chatbot must understand speech as conversation.

In the same way, protocols are arranged in a hierarchy that models different levels of abstraction. The communication substrate just moves bits from one end to the other (and a stream of bits doesn't contain any structure to reveal what it "means") using optical fiber, wire, or radio signals. The level above this understands what the collection of bits means as units of communication. Levels above this manage communication in terms of more complex processes, akin to conversations, including dealing with what happens when the ends become unsynchronized.

5.1 IP

At the lowest level (other than just moving bits) is a protocol that describes how data moves in units, but only simple units. This is called the *Internet Protocol*, or IP. It is responsible for delivering a single chunk of information (in a unit called a *packet*) from a source IP address (the sending computer) to a destination IP address (the receiving computer).

These two computers won't usually be directly connected, so the protocol is responsible not just for moving it from one end of a pipe to the other, but also directing it through a sequence of pipes to get it to the right place. It uses routing tables at each intermediate node to direct each packet so that, at each step, it moves closer to its destination.

The real world analogue is sending physical mail. The post office receives a letter in one location and makes its best effort to deliver it at another. On the way, it passes through sorting offices that each send it on the correct next stage of its journey.

Just like physical mail, IP delivery is best effort and the sender is not informed that the packet has been received[14].

In IP, every packet is treated independently of all of the others, so even packets sent from the same source to the same destination do not necessarily follow the same path, and are not guaranteed to arrive at the destination in the order in which they were sent. This also happens with physical mail.

The pipes that a particular packet might traverse along the stages of its journey do not all necessarily have the same capacities. In particular, a packet might encounter a pipe that can only handle smaller packet sizes (rather like following a short cut while driving and encountering a low bridge). IP has a mechanism for dividing up large packets into small ones and the reassembling them at their destination. This is rather like being able to take a parcel and slicing it into thin segments so that it could be sent looking just like a collection of letters. Unfortunately, IP's mechanism for taking packets apart and reassembling them is simplistic—it relies on each piece recording only how far it was from the beginning of the original packet—which creates opportunities for easy manipulation.

5.2 TCP

Although we managed to use a physical postal service for more than a century, it has obvious drawbacks as a model for data transmission in cyberspace. Two more-abstract protocols are built on top of IP. Traffic using these higher-level protocols is divided into IP packets and transmitted using IP, but they don't need to be aware of how this is done. Instead of seeing pipes as simple bit-moving connectors, they see the pipes as IP traffic movers. These higher-level protocols provide the features missing from IP such as guaranteed delivery and management of the rate of flow.

The most important, and the standard protocol for the Internet, is the *Transmission Control Protocol* (TCP). It is so common that you often see it linked with IP as TCP/IP. TCP provides these features missing from IP:

- Stateful connections between endpoints, creating *sessions* (rather than just one-way single-packet delivery). A session is the

[14]There is actually an even lower level protocol that is responsible for maintaining the relationship between IP addresses and MAC addresses, so that when a new device is connected to a network, the way to reach it can be established. This is invisible to ordinary network traffic.

analogue of a conversation: it has a beginning and an end, and both ends are responsible for managing it while it is in existence.

Setting up a connection requires one end to send a control *segment*[15] to the other end requesting initiation of a session; the other end responds agreeing; and the initiating end then sends a final agreement. Sessions must also be terminated by both ends when their use is complete, using a similar exchange.

- Acknowledgements of packet arrival, so that lost segments can be detected (by timeouts) and sent again by the sender (at the other end of the session).

  Because both ends know that they have a session going, both are expecting traffic from the other end. Once a segment has arrived, the receiver must acknowledge it within a given time (which both ends can agree to adjust). If the sender does not receive the acknowledgement, it assumes the segment has been lost and retransmits it[16].

- Delivery of data to the destination node in the order in which it was sent, using *sequence numbers* carried by each segment. When a segment is delayed or lost, any subsequent segments received by the destination are held until the missing piece arrives. As a result, significant space must be reserved at the destination for reassembling.

- Control of congestion by signalling when the sender should stop sending to allow the receiver to catch up.

- A more sophisticated addressing scheme using both a IP address and a *port number* for both sender and receiver. Each piece of software running on a node and expecting to interact with the outside world uses a particular port. For some common software, the port numbers they use are decided beforehand and widely known—for example, web traffic uses port 80, email uses port 25, and the Domain Name Service uses port 53. Other software chooses a port number each time it is started. These combinations of IP address

[15]The chunks in which data are transmitted by TCP are called segments to distinguish them from the packets transmitted by IP. A segment will typically be divided up into several packets, but TCP does not have to know about this.

[16]There are many optimizations to this process—an acknowledgment is assumed implicitly to be acknowledging all earlier segments as well so not every segment need be explicitly acknowledged.

and port number identify both the node and the particular software running on it so that data comes from and is delivered to the right place.

TCP provides robust communication between two devices on the Internet, but it can be relatively slow if some of the pipes en route are small, when there is congestion, or when the time to travel from one end to the other is long.

5.3 UDP

For settings where the rate of delivery is critical, a second higher-level protocol is also available. It is called the *User Datagram Protocol* (UDP) and it provides the same kind of functionality as IP—best-effort delivery without ordering or error control—but with IP address + port addressing. So UDP trades reliable delivery against fast end-to-end delivery.

UDP is used in applications where delay is more important than the occasional loss of a packet. Applications are typically those that require streaming data primarily in one direction. Examples include Voice-over-IP (VOIP), and video transmission (Skype, Netflix). It is also sometimes used with local networks where high reliability can be assumed (for example, in the Remote Desktop Protocol) since it avoids the overheads of TCP.

5.4 Attacks leveraging protocols

All of these protocols were designed for an environment that was much more trusted, and to deal with systems that had greater performance constraint than most of our systems today. As a result they are conspicuously easy to attack.

Here are some of the common attacks that involve pipes:

- *Eavesdropping*. The entire content of packets and segments is readable unless the payload part is encrypted before it reaches the TCP protocol. Even if the payload is encrypted, all of the details of where the data is flowing and how much is flowing is completely visible within the network.

  This requires access to the network traffic as it is flowing. This can be done via physically tapping traffic (in point-to-point pipes), listening in if the network is wireless, or convincing the network interface hardware to copy all of the traffic to nodes in its local region (a functionality originally useful for detecting problems, but still all too commonly available).

Much of the traffic in cyberspace is unencrypted, so this content can be easily captured and read. For encrypted traffic, the protocol headers are still visible, since the nodes must be able to tell where to send each segment and packet. Traffic analysis reveals how much traffic is flowing between each pair of endpoints, and at what times, The port numbers (in TCP and UDP) are often associated with particular apps, so it is possible to tell what apps are communicating. Sometimes, even when the activity involved multiple apps within a single stream, the packet size profiles can reveal which apps are in use. This can enable the functions of particular computers to be inferred. For example, a web server or print server will have characteristic traffic patterns.

- *Spoofing addresses*. Fake traffic can be sent at will, and legitimate traffic redirected because the IP (and port number) fields in packet and segment headers are not protected. Changing the destination addresses sends traffic to the wrong place, perhaps to a computer that pretends to be the real one. Changing the source address makes it possible to generate malicious traffic but make it seem as if it comes from somewhere else. (This is one of the reasons that attribution of cyberattacks is difficult, since it's straightforward to conceal where traffic originated.)

- *Altering content*. If the payload is not encrypted then it can be altered or replaced without leaving a trace. So the content of some traffic can be replaced undetectably if it is done with care.

  Some parts of the segment headers contain checksums, simple functions derived from the content. These are far from digital hashes; they were designed to catch errors in transmission, not deliberate alterations. If content is altered, the corresponding checksum may have to be recomputed and changed to match, but this is straightforward.

  This flaw in all three protocols is perhaps the main reason why the use of encryption for all traffic is becoming common.

- *Fragmentation attacks* in IP. IP can break large packets into smaller fragments so that data can traverse links that are too thin for typical packets. Each fragment contains a field that describes how far from the beginning of the initial, larger payload it came from. When all of the fragments arrive at the destination (and they may not) IP reassembles them using these fields as offsets. However, these offsets are not checked for plausibility. If an attacker alters these offsets enroute, the reassembly process can cause some parts of the payload to be overwritten by other parts. The resulting packet payload need have no resemblance to the way it started out.

This is not used to corrupt a packet, which could be done directly. Rather the packets carrying the fragments can be made to look innocuous, to a firewall say, while being reassembled into a packet that the firewall would have blocked.

- *Denial of service*. TCP (and, to a lesser extent, UDP) have some state in the interactions between endpoints, and there are attacks that leverage this. The simplest is called a SYN flood attack. Because TCP sessions require a 3-step interaction to set up a session, an attacker can carry out the first step—requesting a session—and ignore the resulting response from the target. The target has to keep track of all of the sessions it is in the process of setting up. When the space allocated for this is full, it can accept no further connection requests, even legitimate ones; but the slots in the table of ongoing connections filled by the attack are never resolved. This is rather like phoning a restaurant, making many reservations under different names for the same time slots, and then never turning up.

  Other denial of service attacks simply keep the target too busy to do anything useful. It is hard for a single attacker's system to generate enough traffic to do this, but an attacker can send traffic to a large number of other systems spoofed to look like it is coming from the target system. These systems then respond to the target all at once[17].

  The receipt of a UDP packets causes the receiving computer to see if there is an app expecting traffic on the corresponding port. If there is not, it sends the equivalent of a "Not known at this address" response. A denial of service attack can send segments destined for unlikely port addresses to many other systems, with their sender address spoofed to be that of the target system to be attacked. The responses then go to the target and overwhelm it.

---

[17] A recent distributed denial of service created 2.3 Terabits of traffic per second aimed at the target.

- *Sequence number prediction and insertion.* An attacker can change the contents of a TCP segment, called a *man in the middle* attack. However, it is harder to insert extra traffic because the successive segments of a session in each direction have increasing sequence numbers. These sequence numbers allow the destination to keep track of which segments have been received and so acknowledge them correctly, but also enable them to be reassembled in the correct order. This makes it difficult to insert extra segments into the traffic (and so take it over in a limited way). If an attacker can guess what the sequence numbers will be (and perhaps stop the real source system from sending by, for example, using a denial of service attack), the attacker can masquerade as the source. Early version of TCP used very simple schemes to generate each initial sequence number but modern implementations take care to make these random. The attack is still possible but requires much more work.

- *Stack fingerprinting.* The official agreements about protocols are underspecified, so that there are places where each particular implementation is free to do whatever is most efficient (or easy). As a result, implementations by different vendors (or for different families of operating systems such as Windows, Linux or Mac) are slightly different. Of course, they must work the same way when it actually matters, but this still leaves room for traces in segment headers that give away which particular implementation of the protocol generated it. This is rather like accents in human speech—we can understand someone with a different accent when they speak (mostly) but their accent reveals something about them unrelated to what they say.

  An attacker can observe traffic and see the traces or consequences of the different implementation choices and so infer what software in running on particular nodes. This may make planning attacks on nodes easier by revealing which make good targets because of known weaknesses.

5.5 Countermeasures

Because the machinery of moving data from place to place was conceived of as a "behind the scenes" and highly technical activity, efficiency was the critical design criterion. Security was hardly thought of.

As a result, there are few countermeasures to defend against these attacks.

A big part of the problem is that pipes have to "speak the same language" so that they can connect a heterogeneous set of nodes and carry a broad mixture of traffic. Any changes to protocols have to be backwards compatible, so that security weaknesses can linger for years, even with determined efforts to improve them.

The main solution to protocol weaknesses is to encrypt the data being moved (but this doesn't prevent traffic analysis), and using protocols that work at a higher level still to look for and correct issues and deficiencies at the TCP level. These protocols, called *application layer protocols*, treat TCP in the same way as TCP treats IP. They assume that there is a way to set up sessions and use them to transfer data (which is what TCP does) but they impose their own checks on data transfers to make sure that it happens correctly.

An important higher-level protocol is a *virtual private network* (VPN) which treats all of the traffic from a node as a single encrypted flow to another, trusted, node where it is unpacked and travels normally to wherever it is going. This protects all of the data as it travels over the VPN and prevents eavesdropping over that part of the journey. Usually, one node is a personal or business computer and the other node is owned by the VPN provider.

Because *all* traffic is mixed together in the VPN, it is much harder to get useful information from traffic analysis or to target one particular session. Of course, this only defends the communication over the section that uses the VPN, so this is only a protection for a particularly untrustworthy part of the Internet. A VPN works rather like ships travelling in a convoy for protection.

Using a VPN has the side-effect of making the traffic appear to originate from the trusted node rather than its real origin. This is used to evade country-specific limits on, say, streaming video services but also for dissidents to use cyberspace without detection or interference from their governments.

Unfortunately, attackers have used VPNs to their advantage by presenting themselves as VPN providers and getting users to sign up to use their VPNs—and then observing the traffic as it passes through the alleged "trusted" node.

Another critical way in which nodes deal with the insecurities inherent in pipes is by using *firewalls*. Firewalls can either be standalone switches or software contained in existing switches. They contain a set of rules about what kinds of traffic can pass; all other traffic is discarded. The rules typically apply at the level of TCP traffic so they can describe which combinations of source and destination IP addresses and ports are acceptable while blocking all others.

Since some ports are strongly associated with particular kinds of traffic, this also means that certain kinds of traffic can be allowed and others disallowed. Unfortunately, although port 80 is reserved for web traffic, so many different actions can be carried out using a web browser, that allowing port 80 traffic through a firewall opens the floodgates to many different kinds of activities.

Some firewalls treat each segment independently and simply check that it satisfies a rule allowing entry. More sophisticated firewalls keep track of sessions (since they can see the segments that set them up) and make sure that all otherwise acceptable traffic also has a session to which it belongs.

Firewalls can exist at all scales—from walling off an entire government or multinational business, down to a single local network.

Firewalls that examine and block traffic carried by an application layer protocol are usually called *intrusion detection systems*. The antivirus and antimalware apps that run on personal computers and phones work in this way. They examine incoming files such as attachments to emails or content of web pages to see if they contain malware. In other words, they wait until traffic from lower levels has been assembled into complete objects, and then examine these objects to see whether they are allowable.

Because they are more aware of the purpose of traffic and because they can afford to spend more time and resources, intrusion detection systems can be more sophisticated in what they detect than firewalls can.

The pipes between two systems can be either point-to-point or involve broadcasting to more than one system. For example, wired networks connect computers to one another using a switch, which makes it harder for an attacker to sit on the network and eavesdrop. Wireless networks, at least potentially, make all transfers visible by every other device on the network, and so eavesdropping is straightforward.

For cell phones, by far the easiest way to intercept traffic is to get the user to install an app with wide access to the phone's apps and capture the traffic while it is still within the phone. A second strategy is to create a *fake cell tower*, convince the phone to connect to it, record the traffic, and pass it on to the real cell tower. Since phones normally connect to the tower with the strongest signal, this only requires the fake tower to be closer than the real tower. Interception of voice calls and text messages sent using standard telephony is straightforward and products to do so can be purchased, and require little expertise to use.

Intercepting wireless network traffic is also straightforward, but depends on the precise network configuration. Network interfaces, both switched and wireless, are capable of broadcasting to all nodes in their local region, and this can often be turned on surreptitiously by an attacker.

Finally, pipes can be physically broken (often by digging in the wrong place) and when this happens it can cause large-scale chaos.

Chapter 6

Configuration security

The nodes and pipes that make up cyberspace must be able to talk to each other so the rules for interaction, the protocols such as IP and TCP, make this possible. But the system of nodes and pipes must all work together as a whole to cope with a changing environment—nodes and pipes disappearing or appearing, transient effects like congestion—even though the system is decentralized and owned and managed by many different entities. This requires a way to manage the system in its entirety that works even though it's decentralized.

This organization reflects the fact that the original Arpanet was designed to survive a nuclear attack, and so was designed to be decentralized. All of the basic operation of the Internet has inherited this decentralized design, and so there are few points where Internet properties can be controlled directly.

6.1 Internet Control Message Protocol

The Internet must have a way of maintaining the routing tables at each switch, and discovering when something has gone wrong and dealing with it. The mechanism uses the *Internet Control Message Protocol* (ICMP). This protocol runs at the same level as IP, and can be thought of as a parallel protocol. Users don't have to be aware of it because it takes care of configuration behind the scenes, but it contains vulnerabilities that can be used for attacks, so it has to be understood for a complete understanding of cybersecurity.

Cyberspace is a dynamic system, particularly because systems near the edges come and go (as people turn their computers on and off, for example). Other more important systems can also come and go, as they are taken offline for maintenance or because they crash (although mainstream providers design their systems so that maintenance can be done seamlessly). Switches can also turn on and off as they, for example, reboot after being updated,

There has to be a way of responding to these changes while keeping operations running as normally as possible, and that is what ICMP does. It has three main functions:

- Helping new endpoints find a switch that can route traffic to wherever they want to send it. Whenever a new computer joins a network it broadcasts a request for a *router* (a switch containing a routing table). A nearby router will respond, and the new computer will then know where to send outgoing traffic to get the traffic started on its way. ICMP contains the functionality to carry out the stages of this process.
- Discovering when a pipe or switch is not functioning, and working out new routes to get around the problem. This is done by including a "time to live" counter in an ICMP packet which is decremented each time it travels from one node to the next. If the counter reaches zero, a message is sent back to the originating node to say so. A node can use this functionality to probe a pathway to see if it is operating properly.

  Although ICMP is a low-level protocol, this part of its functionality is available at the application level (and so to computer users) via two apps called *ping* which, when given an IP address, reports whether it exists on the Internet at the moment, and *traceroute* which, given an IP address, reports all of the intermediate nodes on the current path to it (if it exists). These apps were designed for debugging network connection problems, but they also reveal a lot about how systems are connected that can be used by an attacker for surveillance.
- Handling congestion by allowing nodes to request that traffic coming to them should be rerouted another way. This allows switches to deal with temporary, local issues. However, it can (and has) been used to cause switches to reroute all of their traffic through, say, another country where it can be analyzed before onward transmission.

In a pattern we've seen before, the ICMP protocol was designed to manage the operations of networks under the assumption that ordinary users would not care about how it worked nor have access to it. It therefore does not contain any security protections.

6.2 Domain Name Service

The other behind-the-scenes system in cyberspace is the *Domain Name Service* (DNS) which takes care of the mapping between human-readable computer names and IP addresses. It is a distributed set of data tables recording the current mappings.

Traffic to and from DNS servers uses UDP, but this traffic is invisible to users—it is embedded in web browsers, mailers, and other applications that need to resolve names to IP addresses.

A *domain* is a part of the name space that exists because computer names have a hierarchical structure. For example, ".uk" is the domain of all computers in the U.K., "gov.uk" is the set of all government computers in the U.K., and "data.gov.uk" is the domain of systems that hold open-source data from the U.K government.

In each subregion of the Internet, at least one computer, the local DNS server, will hold mappings of names to IP addresses, usually those that are relevant to whatever the local computers are doing. For example, these local servers might hold the mappings for all of the computers that belong to the same business or institution, local or remote, because the local computers are likely to communicate with them; and also for popular web sites for news or social media that users on the subnetwork typically use.

If a device asks for a mapping that is not held by its local DNS server, this server sends the request to a higher-level DNS server. If that server knows the answer, it will return the mapping, which will be sent back to the original requesting computer, and also stored in the local DNS server. This last step is done because once one local computer requests a mapping, it is likely that it or one of the other local computers will also want to know that mapping in the near future. If the DNS server runs out of storage space for mapping, it will delete the mapping that has been used the least in the recent past.

If the higher-level server doesn't know the mapping either it will send a request to the next highest level, or perhaps skip some levels and go straight to the top level DNS server. This top level DNS server knows how to find the DNS servers for ".com" and ".uk" and all of the highest level domains.

Some DNS servers in the hierarchy are *authoritative* for a given domain and will never delete the mapping for that domain. Thus there is an authoritative DNS server for ".uk" and other DNS servers can always find a subdomain within ".uk" by asking for it from the server that is authoritative for each subdomain. In other words, authoritative DNS servers hold *the* definitive mapping between a domain name and an IP address while lower-level DNS servers hold *copies* of such a mapping.

The DNS system is complicated by the ability to define *zones* which crosscut the hierarchical structure of domains and subdomains. A business might have domains in different countries, for example, *fredmart.com*, *fredmart.co.uk*, and *fredmart.com.au* and might want to redirect web site visitors who visit any of their web sites to the version appropriate for their country. They can set up an authoritative DNS zone server for all of the *fredmart.** domains to make this happen.

There doesn't have to be a 1-to-1 mapping between names and IP addresses. It is common to map a name to more than one IP address as a way of load balancing, so that traffic gets spread to different physical systems. This ability also provides a way to deal with denial of service attacks by seamlessly providing a way to distribute high-volume incoming traffic, even perhaps diverting traffic from the locations participating in the attack so that attack traffic interferes only with other attack traffic. Popular web sites can also add extra servers at busy times in a transparent way.

Although the DNS is a low-level system it also has an application-layer interface using the *whois* command. This command will provide the current mapping between a domain name and an IP address as well as providing information on who owns the address and what its authoritative name server is.

The DNS system has significant vulnerabilities. An attacker who can cause a DNS server to pass a request on to a higher-level server and then simulate a false response from that server can cause a false entry to be included in that DNS server's mapping table. Subsequent requests for a mapping by ordinary apps will be invisibly diverted to the wrong location.

DNS servers are also vulnerable to denial of service attacks. An attacker can generate many requests for mappings for non-existent domain names. The mappings for them will not, of course, be known and so will trigger a cascade of requests to higher-level servers, none of which will know the mapping, and so on.

The weaknesses of DNS have been partly addressed by allowing requests and responses to be encrypted. This makes it harder for an attacker to insert false content because it wouldn't be encrypted properly. However, older systems must still be served so this won't help much until encryption of domain name requests and responses becomes obligatory.

6.3 Switch vulnerabilities

Switches are just computer systems configured so that they can handle lots of data transfers, and so they can operate without direct human intervention. They have IP addresses of their own, and they run standard system software. Switches are points of special vulnerability because they are involved in all aspects of moving data. Being able to manipulate switches creates unprecedented power to attack and disrupt.

Switches have the same vulnerabilities as any other nodes. In particular, they must be able to install updates, and so they have their own connection to the outside world—and a management account that allows them to be accessed remotely, even though their normal operation doesn't require a human operator. Such an account has to be one with high privilege because the entire switch environment has to be able to be manipulated remotely—many switches are in places where direct access to them is difficult or impossible. This means that switches can be manipulated maliciously from anywhere else on the Internet—changing their configurations, or even causing them to turn off.

Switches are also a vulnerability because, of necessity, they see all traffic that flows through them. If they can be corrupted, then the data for traffic analysis can be collected without having to attack pipes.

Switches are also vulnerable to physical attack because they are often physically and geographically isolated. It is possible to physically access many switches and simply remove or destroy them. Their environments can also be altered, say by heating them beyond their thermal rating, and so prevent them from operating.

6.4 Mounting an attack

We have seen that some attacks are easy to mount because they rely on the unwitting collaboration of a user on the system to be attacked (especially phishing and spearphishing). Such attacks can also be a stepping stone to attacks on other systems.

Attacks that cannot rely on users helping them must take a different approach. Although it's possible that such attacks are aimed at targets of opportunity, they are more often targeted at a particular computer belonging to a particular organization. For example, sensitive data that an attacker might want to exfiltrate might be stored in a contracts database. The target computer may be deep within the network environment of that organization. protected by multiple layers of firewalls and perhaps an intrusion detection system.

An attack like this usually requires a strategic approach that involves the following steps:

- *Surveillance.* There is no straightforward way to identify the particular computer target. Computers that are not Internet-facing do not have names that are spread to DNS servers, so it is difficult to find IP addresses, and even these only reveal the subnetwork where the computer is present, not how to get to it.

  An attack may have to work its way inwards from Internet-facing computers to their neighbors internally and then on to the intended target (so one attack actually requires multiple other attacks as stepping stones). Obviously, the set of stepping stones should pass through the most vulnerable computers, perhaps even at the expense of taking a longer path.

  So another facet of surveillance is identifying computers that might be vulnerable. This is one reason why stack fingerprinting is useful—it allows the attacker to identify what versions of software are being used by a computer. Knowing a list of vulnerabilities, an attacker can assess how easy a target each visible system is[18].

  Surveillance may also include looking at the traffic originating from and destined to each computer. This may reveal what its purpose is: is it print server? A database front end? Many devices that don't look like computers, for example, printers or IoT devices, may also be open to attack, and provide a foothold for further steps. These devices are fully functional computers even though we tend to be blind to this because they appear to have a single function.

[18] Vulnerabilities are widely known because system administrators have to know about them to fix them. A public list is maintained at `cve.mitre.org/`.

Tools such as ping and traceroute can also be used to explore the environment inside an organization and try to work out IP addresses and structures. For example, not every possible IP address on a subnetwork will necessarily have a device associated with it, and there's no point attacking non-existent devices.

The Time to Live feature in ICMP can be used to probe firewalls. If a packet is sent with a Time to Live one greater than the number of steps to the firewall, there will be no response if the firewall blocked it, but a bounce if it made it through the firewall (but then ran out of life). An attacker can use this to see whether traffic to a particular address and port combination is allowed through the firewall.

This probing can also provide information about what the destination computer is like. For example, if port 80 traffic (used for web browsing) is blocked, then it's unlikely that the destination computer has ordinary users.

- *Attack planning*. Once an attacker has an understanding of the systems and networks that lie between the outside world and the target system, an attack can be designed. This requires choosing attacks for each system on the required path, as well as fallbacks should an attack not work as anticipated. An attacker must have a repertoire of attack techniques and tools and an awareness of which systems are vulnerable to each.

- *Actual attack*. The actual attack requires carrying out each of the planned steps in sequence, Some phases of the attack might involve probing to see whether a particular attack will be successful. If not, another approach must be used instead.

  Because an attack usually requires a chain of intrusions into nodes along the way, an attacker will usually insert mechanisms into systems that have been attacked to allow future access without having to use an attack again. These are called *backdoors* and might work by creating a new administrative user account, or by reconfiguring an interface so that it is less secure, and so it can be used again without effort.

- *Cleaning up traces*. All computer systems and switches keep log files that record the actions that take place. These can be at all levels of granularity from TCP sessions up to commands issued by each logged-on user. The main purpose of these log files is to be able to work out what happened when something goes wrong with a node or pipe. However, they capture the activities of innocents and attackers alike.

An attacker is therefore motivated to remove log files or change them so as to leave no traces, either to conceal that there has been an attack or to prevent it being attributed correctly. Removing log files leaves no traces of the activity being logged, but the fact that they have been deleted is itself suspicious. Altering log files is preferable but it requires more sophistication, and it is easy for an attacker to miss a trace left in some log file when there are so many of them.

- *Leaving a backdoor*. After all of the effort required for an attacker, it is natural to change the target system so that a repeated attack is much easier, if required. We have already mentioned leaving systems along the way easier to reuse, but making the target system easier to re-attack is more useful. (Of course, altering the target system to make this possible risks creating extra traces. So for exfiltration and other covert kinds of attacks, it might be more appropriate to leave the system looking completely intact.)

6.5 Defending against attacks

Defenders could find the traces of an attack in progress, or even one in preparation if they looked for them. But the reality is that most systems are under constant low-level attack, almost all of which are defeated by firewalls and intrusion detection systems. These systems can be configured to set off an alarm to a human if an attack looks more powerful than the typical background, but setting the threshold for these alarms turns out to be difficult. Set them too high, and important attacks are missed; set them too low and there are constant false alarms. When there are too many false alarms, humans learn to ignore them. Social engineering can be used by attackers to set off alarms deliberately until they get ignored; and then the real attack can be carried out.

The most difficult case is what are called *Advanced Persistent Threats* (APTs), which are attack preparations that are potentially serious but happen really infrequently, perhaps weeks or months apart. These are widely believed to originate from states, patiently accumulating information about potential targets. They are difficult to notice because each individual step seems fairly inconsequential and the sequence is hard to detect because of the time gaps between them.

It's natural to ask why computer systems aren't all as well defended as they can be. We touched on this in Chapter 4—the behavior of pipes is practically impossible to update; and nodes are updated on a regular schedule, typically monthly or biweekly.

However, many systems are not even updated to the current best level because of a concern that updates will break the ongoing operation of nodes. A business may not be able to afford the downtime that could be triggered by an update that interacts poorly with their core revenue-generating systems. Such businesses may trial each update in a walled-off environment to judge if its deployment will cause issues, and this process takes time. There may therefore be organizations whose update state lags behind what it could be by months, and these yet-to-be-updated systems are natural targets for attackers.

There are also a surprisingly large number of systems that are unattended, and have been forgotten by their owners. Such systems may never be updated, and so represent serious vulnerabilities for their organizations. Some systems, notably Internet of Things devices and routers are impossible (or extremely time consuming) to update.

There are resource implications to keeping a large set of computers up to date, and so many organizations also find themselves forced to triage to keep their most critical systems best protected, and live with some level of risk for less-important systems. Organizations can get a perspective on their level of risk using *attack graphs*. These represent how easy Internet-facing systems are to attack, and then transitively assess how easy the systems one step away from the Internet are to attack, and so on, based on the versions of all of the software that they have installed. An attack graph can indicate to systems staff where the most cost-effective interventions can be made to protect the overall system. (An attack graph is also what an attacker would most like to have, and is attempting to construct implicitly by surveillance.)

Defenders can also keep track of incidents that are indicators of surveillance or early-stage attacks. One way to do this is to look at the logs which record many activities on systems. This is not done as often as it probably should be because logs are very detailed (and grow quickly) so it is hard to pick out anomalous activities that might signal probing or attack preparation. Tools to partially automate this step are becoming available. Nevertheless it is still true that logs are often only looked at afterwards to work out how an attack happened and to attribute it, rather than as a real-time defensive tool.

Some organizations set up systems called *honeypots*, systems that are designed to look like attractive targets but actually contain nothing of real value. Honeypots can be watched by defenders more carefully than other computers, so that they act as tripwires, notifying defenders of a potential attack. Their other advantage is that a defender can examine the attacker's strategy as it is revealed in the attack on the honeypot, and so see how that attacker might approach other computers in the organization.

In organizations where security is paramount, computers may have their entire software and systems reloaded every day. This means that an attacker has a very limited time to exploit a successful attack, and cannot leave behind a backdoor for further attacks.

6.6 Recovery

When a successful attack has taken place, a defender has to carry out a number of tasks to recover, and perhaps to deal with the attacker legally.

Obviously, the first phase is damage control. Something will have happened that causes the defender to realize that an attack has taken place. The defender must try to work out what the damage to the organization has been and remediate it if possible. The defender will then need to work out how the attack was carried out. This will often require careful analysis of logs to find the traces of the attack.

The second phase is to fill the holes that allowed the attack to happen. This might be straightforward if the weakness was the lack of updates but might be more difficult if the attack used a zero-day mechanism because the defender cannot easily tell exactly what happened.

The third phase is to work out how (if possible) to attribute the attack, and perhaps take retribution. Attribution is difficult in principle because of the ease with which the source of an attack can be hidden behind spoofed source addresses. In practice, however, attackers are careless and sometimes unaware of what is collected about activities, and it may be possible to infer, with some assurance, who carried out an attack. Curiously developers seem unable to resist the temptation to sign their work when they build malware, and this has often helped with attribution.

When it comes to retribution, organizations such as businesses can try to involve law enforcement. This tends not to work well because police forces do not have large numbers of employees with the skills necessary to understand an attack, and neither do courts.

Governments tend to be proactive about attribution because they need to understand how attacks fit into larger international relationships, and may resort to active cyber operations as a form of cyber deterrence. This is a rapidly developing area.

Chapter 7
Application security

Most applications communicate by using the protocols we have already discussed, notably TCP/IP. However, there are protocols that allow particular apps running on different systems to communicate with one another at a higher level of abstraction, that is, treating the entire pipe system as a communication mechanism without having to pay attention to the details of how it works.

Working at this higher level of abstraction also makes it easier to see that certain types of traffic flows should not be allowed. This done by *intrusion detection systems*, which are analogous to firewalls, but operate within individual nodes, rather than in the pipes. Because of where they are, they can better see which patterns of traffic flows make sense (to and from running apps) and which do not, and can block those that don't make sense. Because they run within nodes they also have the resources to examine traffic in more detail and with longer deadlines, so they can do a more thorough job.

We now consider two of these application-layer systems.

7.1 Email

Sending an email requires the sender's email client to send the message to its local email server. This email server must then transfer the message to an email server at the destination, and this server passes it on to the email client belonging to the receiver. The communication mechanism between the two servers uses an application-layer protocol called the *Simple Mail Transfer Protocol* (SMTP) . Email was an early use of a network, and this protocol dates back to the days of trusted networks. Security was not a consideration when it was defined, because email was a way to move simple text from a sender to a receiver.

Email has since been overloaded with many other functionalities such as moving arbitrary files as attachments, or organizing meetings. Security mechanisms, minimal to start with, have not kept up with all this new functionality.

The nature of email means that security is inherently difficult. Email servers must be easily discoverable; otherwise a sending email server wouldn't know where to send each email. (It is actually a bit more complex and subtle: an email sent to an address of the form jsmith@corporation.com will actually be sent to a domain such as mailserver.corporation.com, so more than just DNS is involved.) This means that at least one computer for any organization must be visible to the outside world, providing an unavoidable starting point for an attacker to learn about the organization's internal computer system setup.

By design, emails are supposed to get from one email server to another, so this kind of traffic cannot be blocked by firewalls. Intrusion detection systems must do the work of deciding whether an email should be allowed to pass. The email system is open in the sense that anyone is allowed to send an email to anyone else whose email they know or can guess, so IDS rules don't usually block emails based on where they come from. As the problem of spam shows, this weakness causes annoyances for users, and spam detection systems only do a moderately good job of dealing with it.

Email contents are not usually encrypted, so the contents can be read by any system or switch through which an email passes. There have been many attempts to address this issue, all of which have been too much work to have received much uptake. Public key cryptosystems do, of course, work. If you know the public key of the person you are sending an email to you can encrypt the contents with that key and the receiver can decrypt it, confident that it couldn't have been read in transit. Doing this requires knowing the receiver's email address and public key which could be done by the sender on a one-off basis, but would require a major redesign of the entire email system to automate.

Originally emails were simply text, but email today can contain attachments in a wide variety of formats. Many of these attachments can legitimately contain executable code. For example, a Microsoft Word document can contain executable macros that have an effect on how they are displayed, even though a Word document looks like it ought to be just a container for text and images. This makes it possible to construct attacks in which the executable code in an attachment installs malware on the system where the email is opened, which is the basic mechanism of phishing. Although antivirus and antimalware software routinely scan these attachments, it is difficult to tell when or if a particular one contains malware.

Mail clients contribute to the problem by making it easy to click on attachments to open them (a convenience but also a weakness) and allowing the displayed file name to be chosen by the sender so that, say, an executable file can be made to look like a text document. Email messages can also contain links that can be clicked within the mail client allowing the email recipient to be directed anywhere and, again, the apparent destination of the link text need not be where the link actually points.

The entire email system contains no checks on the content of messages or even on the identity of senders. It is straightforward to create a false From: address to conceal the real identity of the sender. Malicious emails do not trivially reveal who sent them.

Email messages keep track of which systems they have passed through, although this information is not usually visible in the email headers without special effort. This has some advantages: it allows spoofed senders to be tracked back to a system (close to) to where the message originated. However, it also impacts the privacy of ordinary senders who might not want it known where they are posting from.

It is convenient to be able to send emails from somewhere other than your home location, for example, at hotels, or on planes and trains. Email servers were often configured to allow anyone connected to them to send emails, an approach called an open relay. Sadly spammers started using these systems to send large amounts of spam. Increasingly, to send email you must first connect to your home system or otherwise authenticate yourself to your own local email server.

There is no flow control in email systems so it is possible to overwhelm a particular system just by sending it a lot of email in a short period of time. Email communication is not synchronous, so emails can pile up at a destination server waiting for someone to download them to a client. This storage of data can easily use up large amounts of space and accidentally or deliberately cripple an email server.

Email is a best-effort communication mechanism. Some email systems provide a way to notify the sender that a particular message has been read, but this is so often misused—for example, by spammers to determine if a particular email address is live—that most users turns it off even on systems that provide the functionality. However, most people assume that email is reliable, and this can sometimes lead to problems and miscommunications.

7.2 Web traffic

The World Wide Web is the way that most people interact with cyberspace. It is therefore a major target for malicious activity.

The Web started as an information delivery platform. Using a browser, anyone could enter a URL, and get a web page of content. Web pages contain links so that a user can move from one page to a related one by clicking on the link. Web search engines make it possible to use search terms to get a list of ranked relevant pages when URLs are not known.

The application-level protocol used for Web traffic is the *Hypertext Transfer Protocol* (HTTP) which defines how browsers and web servers interact. Web traffic uses port 80, that is, browsers send requests to web servers using the destination IP address and port 80, and get results back via the same port number. Because web access is so ubiquitous, almost all systems allow access to this port, that is, firewalls hardly ever block it.

Increasingly, HTTP has been replaced by HTTPS (HTTP Secure), a version of the protocol that encrypts the traffic in both directions between a browser and a web site. This protects the content but not, of course, the URLs being visited. Using HTTPS requires web sites to set up the machinery for public-key encryption, which has overheads and can be too complex for the web sites of smaller or less-sophisticated organizations. As a result, HTTPS is still not widely used, although it has become common for the largest and most popular web servers. Some browser developers have tried to put pressure on for wider adoption of HTTPS, by threatening to require it, but so far this has not been successful. HTTPS traffic uses port 443, instead of port 80, so this port is almost inevitably also left open by firewalls.

The browser-server interface was designed to do something relatively benign and straightforward (move web pages from servers to browsers) but the interface now has much greater functionality: browsers can send content to web servers, and web servers can send code to browsers which is executed by the browser, and so local to the user. This allows users to play games or view video in their browsers. This new functionality resulted in changes in the definition of the HTTP protocol not all of which were well thought out or designed, and for which (as usual) convenience trumped security.

Browsers can transmit data to web servers. Sometimes this is simply the content of a form, often as the way to authenticate to the web server using a user name and password. But browsers also allow a user to transfer money between bank accounts, buy products, or borrow ebooks from a library. The same mechanism allows users to post content to a web site which is then visible to others on the same web site. For example, user can upload videos, status updates, blog posts, and share their activities such as what they have bought.

As well as the explicit communication between browser and web server, there is a great deal of implicit communication aimed at tracking user behavior (for example, using *cookies*). Because many users are interacting with multiple web sites at once (perhaps using separate tabs) there is an opportunity for interactions with one web site to leak information about the interactions with another web site.

A mechanism that was designed to handle simple requests (upward) and static data (downward) has become a channel for high-volume two-way transfer of many different kinds of data, as well as executable code.

Web pages contain Hypertext Markup Language (HTML) which describes what each part of a web page is, rather than its exact formatting. For example, HTML tags label paragraphs, headings, lists and so on. The browser can use this information about what kind each part is to decide how to display it, knowing the screen size and resolution of the device that will display it.

The major vulnerabilities of web traffic are made possible because parts of a web page that are code or user-generated content are labelled with tags as well. Thus text, inside each of these parts, becomes executable code when it reaches either the browser or web server. Checking whether this executable code is reasonable and safe is difficult, and is often not even attempted.

Three major attacks exploit this ability for data to become executed code. The first is called a *cross-site request forgery*. An attacker uses their own web site to make a malicious request to another web server via an unwitting user's browser. For example, suppose that a user is logged into her bank account at a bank's web site, and simultaneously has another tab open at the attacker's site. The attacker can cause the user's web browser to download a malicious request (say to transfer money) and then pass it up to the bank's web site. The bank knows that the user is already authenticated, and doesn't realize that the request does not come from the user's actions, but indirectly from the other site. The user may not even realize that anything has happened.

The second is called a *cross-site scripting* attack and works in an inverse way. An attacker inserts HTML-formatted content on a web site that allows users to upload content—for example, a social media site might allow a user to describe their interests. Inside this uploaded content is an HTML section that contains executable code. Whenever a user downloads the content, this code is executed on their computer and can do anything that executing code in the browser is allowed to do. The downloaded malicious code may do something bad directly, but can also cause other malicious code to be downloaded.

Both of these attacks seem like they should be easy to prevent, but they interact with core mechanisms of HTML and browsers: how executable code is handled, how authentication and cookies work, and the ability to interact with more than one web site at a time. Several attempts have been made to prevent these attacks, and it is an evolving field, but a solution requires agreements between all browsers and all web sites about exactly what needs to be done. In practice, browsers have attempted different ways to try and solve the underlying problems without needing web sites to be changed (since that will never happen across the whole world).

The third attack is called *SQL injection* and is perhaps the most popular single cyberattack mechanism. Web sites often create pages on the fly when they receive a request. For example, if you visit your bank, you must first complete a form that provides your account identification and password; but then the bank displays a page with your balances and other individual details. This page is, of course, different for everyone.

The data that populates this page comes from a database which has records for each customer. Databases are specialized data stores which provide information in response to requests expressed as queries: "show me the balance of the individual with this account number", "show me the account numbers of all the people at this branch", and so on. These queries are expressed in a structured way using a query language, of which *SQL, Structured Query Language*, is one of the most popular.

When you enter information into a form in your web browser ("what is the balance in account xxxx") it becomes an SQL query which is passed to the web server, and the web server passes it on to the database. When the database responds to the web server, the server turns the response into HTML and sends it to the browser for display.

The problem is that the web browser and/or web server do not (and in some cases, cannot) validate that the text entered into the form is a safe SQL query. Thus a knowledgeable user can enter information into the form that carries out malicious actions in the database, up to and including deleting the entire contents.

Several countermeasures to SQL injection attacks are known, mostly based on working harder to validate inputs before they reach the database. However, these rely on correctly configuring both the web server and database. Many web servers (and databases) are built for small and medium size businesses who may not be aware of the issues, and so misconfigurations that allow these attacks are still common.

These attacks can act synergistically. The cross-site request attack allows a malicious web site to create a request to another web site. This request can be an SQL injection attack on the database that lives behind that other web site.

The code used to run web servers may also have vulnerabilities that are known to potential attackers. Even well-known web servers that have been in existence for a long time contain errors. For example, the Apache web server issues roughly 10 security fixes each year. A sophisticated attacker may be able to attack such web sites directly, that is, without relying on a browser as an intermediary.

Many of the attacks that allow malicious actors to get into individual's social media accounts or bank accounts use phishing or spear phishing to insert malware on the individual's computer or phone. Such malware can watch when the individual logs in to their account, capture their passwords, and then quietly log in behind the scenes; and then post content to social media, send email spam, or transfer funds out of their bank account. Users can protect against these attacks by using antivirus and antimalware software that detects malware like this before or after it gets onto the device.

Attacks using web server mechanisms are harder to detect because they are mostly transient—the attack happens in a moment, creates no sign that anything is happening at the time, and leaves few traces.

7.3 Blockchains

Blockchains are an application-level system that relies on many of the cryptographic ideas we discussed in Chapter 3. Although the long-term usefulness of blockchains remains an open question, they create a complex security environment which is worth exploring.

Recall that a digital hash is a function that maps strings of characters to a fixed-size representation (often 32 bytes in this context) in a way that cannot be practically reversed, and in which two similar input strings are mapped to completely different fixed-size representations. They represent summaries of the content of the input strings.

Recall also that a public-key cryptosystem uses a pair of a private and a public key to allow two parties to communicate using encryption without having to exchange keys. Such systems allow anyone to send content to the holder of the private key, and for the holder of the private key to digitally sign anything in a non-repudiable way.

A blockchain is a distributed transaction-recording system that is intended to be a permanent record of the transaction history. It is made up of blocks, each of which contains a timestamp and the record of one or more transactions, connected in a chain. Each block contains a digital hash of its predecessor so that the previous block cannot, in practice, be altered[19]. They are designed to be *tamper evident*, so that attempts to alter the blocks or the chain are visible, and *tamper resistant* so that it is hard for malicious actors to change or replace transactions using the mechanisms of the blockchain.

A user who wants to add a transaction to the blockchain must own one or more private-public key pairs. Activity on the blockchain is associated to the identity that holds each pair of keys, but there is no necessary link from the identity to the particular individual who owns them. When a user wants to add a transaction to the blockchain, it is broadcast to participating computers (nodes), who then compete to be the one allowed to add the transaction to the chain. The node that wins is called the publisher node. Each blockchain system must have some way to reward publisher nodes so that they have an incentive to compete. This could be as simple as a transaction fee.

There are a number of mechanisms for this competition that different designs for blockchains use. All are designed to prevent one node or a set of nodes from colluding to always win, and therefore be able to corrupt the process of inserting new blocks into the blockchain.

The first mechanism is called *proof of work*. Here, potential publisher nodes must solve puzzles that are computationally expensive to

[19]Because to do so would require finding a version of the previous block that would hash to the same value, which digital hashing is designed to make computationally prohibitive

solve, but for which the correctness of eventual solutions can be easily checked. A set of nodes that wanted to collude would have to devote huge resources to always being the first to solve puzzles, and so to gain control of the blockchain. The first node that solves the puzzle gets to add the next block to the chain, and propagate the extended chain to all of the nodes involved in the blockchain.

It can happen that multiple nodes solve the puzzle simultaneously and each add a (potentially different) block to the blockchain[20]. There are now multiple versions of the chain in existence at the same time. Nodes that receive chain updates that are inconsistent prefer the longest one, and the transactions from the other chains are put back into a transaction pool. Thus the inconsistent versions are gradually brought into agreement.

There are two major problems with proof of work approaches to blockchains. First, they are an ecological disaster, since many nodes compete to become publishers by carrying out expensive calculations which consume a great deal of electricity[21]. Successful competition typically requires substantial computing power. As might be expected, nodes have tended to form cartels to reduce competition; but this weakens the resistance of the whole approach to collusion.

Second, because the chains can be inconsistent for a while, it is safer to wait until multiple blocks have been added to the chain before concluding that your particular transaction has been permanently added to the chain—as a rule of thumb, wait for six further blocks. Since some blockchains add a block only once every 10 minutes, this introduces substantial latency into successfully concluding transactions.

A second mechanism is called *proof of stake*, in which the winning node, the one that gets to publish, is randomly chosen in proportion to a stake, typically monetary, that they have put into the system. The idea is that nodes that have a stake in correct operation of the blockchain are motivated to be hard to suborn. The process of randomly choosing the winning node is carefully designed to make it as difficult as possible to corrupt the process. Careful design of the process for becoming a publisher node can make it highly unlikely[22] that more than one node can think that it is the winner in a particular round, so these blockchains do

[20] This is called a *fork* but this term has also been overloaded to mean a branch in the chain caused by, for example, a change in the encryption scheme.

[21] It is reported that Bitcoin, one major blockchain causes as much electricity consumption as a smaller European country such as Belgium or Denmark.

[22] 1 in a billion for Algorand.

not create inconsistent versions, removing the need for an inconsistency removing step.

Although proof of stake approaches do not require the huge amounts of computation required for proof of work, they still have performance costs. For example, long-standing blockchains can have chain sizes as big as 200GB (and of course they have to grow), so storing and moving the chain is expensive.

Blockchains can be permissionless, so that they are truly distributed and users and nodes need have only minimal trust in the other entities involved. Much of the expectation of large-scale systemic change because of the use of blockchains comes from the decentralization, in which transactions no longer require a trusted intermediary (who usually takes a small fee for the service). Permissionless blockchains, it is claimed, destroy many forms of intermediaries or gatekeepers.

Blockchains can also be permissioned, for example, because they operate entirely within a large organization. This reduces the friction of some of the blockchain processes because there can be greater trust between the entities.

One of the most visible applications of blockchains has been as ways to implement *cryptocurrencies*, of which Bitcoin is probably the best known. Here the blockchain is used to record all financial transactions, so that there is a complete record of who has owned each unit of value (e.g., Bitcoin). Value enters the blockchain from the wider financial system; once in the blockchain it can move from ownership by one identity to another; and eventually it can be moved back into the wider financial system. Because identities are known only from by encryption keys, users can transfer value anonymously. This, obviously, creates pathways for criminals and money launderers to move value around in ways that cannot be tied to them. Note that all transactions are visible; it's just that the identities of the individuals who made them that are concealed. Cryptocurrencies therefore function like "Swiss bank accounts" because you can't tell who owns the value.

The ownership of units of value is associated with the private key that was used to acquire ownership of them. Most of the high-profile hacks that have "stolen" cryptocurrency have really been illicitly gained access to private keys, enabling units of value to be transferred to someone else. These private keys are sometimes called digital wallets, because ownership of a private key is equivalent to ownership of value units recorded in the blockchain.

A blockchain used for cryptocurrency has a natural way to reward publisher nodes, by giving them cryptocurrency as a reward. It took nations many centuries to understand how to keep their currencies stable, and cryptocurrencies have not yet solved this problem. Their prices relative to real-world currencies have oscillated wildly, so that the incentives for nodes to become publisher nodes have also oscillated, making it hard for potential publisher nodes to plan their resourcing.

There are many cybersecurity vulnerabilities associated with blockchains. These include:

- Attacks directed at the distributed mechanisms for choosing a publisher node, which are designed to be fair and random. These include collusion to allow a particular node to be selected as a publisher node, and so-called 51% attacks, where more than half of the potential publisher nodes are corrupted so that they can create their own, competing version of the blockchain, with its own transactions (adding extra, or deleting some of the true ones) that is longer than the true one and so gets adopted as the authoritative one.

- Attacks that exploit the fact that transactions that have not yet been added to the blockchain are just data, and can be discarded, changed, or misdirected.

- Attacks on the (distributed) governance structure that balances the power of software developers, the set of potential publisher nodes, and the users themselves. Each of these groups has some leverage over the others, and enough malicious members of any one of them can destroy or corrupt the blockchain. Poorly written software can also cause issues, even if it is only created by accident. For example, software errors in the code used to move value from the financial system to a cryptocurrency blockchain have led to value totally disappearing—it is, so to speak, in the blockchain but in a way that prevents it ever being transferred or extracted.

- Botnet-based attacks to gain the rewards of becoming a publisher node. Unscrupulous actors have used botnets to carry out the required calculations, even to the extent of using the browsers of unsuspecting users as computational engines. Since such a publisher node does not have to pay for electricity or equipment, the economic cost-benefit of proof or work is more attractive.

- Attacks that exfiltrate private keys, that is, steal digital wallets. Since private keys must be stored somewhere (they are too long to type in each time), they represent an attractive target. Malware can either search the computers of cryptocurrency users looking for strings of the right length to be a private key, or watch the cut-and-paste buffer when the user is transferring value to or from the blockchain; and then steal that string.
- Denial of Service attacks carried out against the nodes of the blockchain network preventing them from seeing new transactions, and so not competing to become the publisher node for them. This can disrupt the entire operation of the blockchain, or skew the competition by eliminating some potential publisher nodes from the competition in a round.

Blockchains rely on the cryptographic protocols they use being, and remaining, strong enough. There have been several examples of holes in either the algorithms themselves, or in how they were implemented. They face a special problem because they are intended to be a permanent record. It is therefore hard to change the fundamentals of the way they work without creating a break from the previous state of the blockchain; and sharp changes like this often create mistakes and vulnerabilities.

Chapter 8
Summary

Most of the challenges of cybersecurity can be traced back, ultimately, to the origins of the Internet in the research network called Arpanet. Arpanet connected small numbers of universities and U.S. government research labs so its users were trusted and accountable. Arpanet was also designed to survive a nuclear attack and so its operations and control were distributed by design.

Arpanet was never intended to be the foundation for a global network that now connects half the world's population, and many billions of devices. Today's Internet reaches into every nation on earth, and has capabilities that could not have been imagined by the original designers of Arpanet, including the World Wide Web, ecommerce, online banking and financial services, social network platforms, distributed control of pipelines and electricity grids, and remotely controlled lightbulbs and thermostats. As well as being extremely large, the Internet is dynamic because devices come and go as they are turned on and off or fail, and because data flows over the momentarily best path between each source and destination.

The distributed nature of Arpanet meant that, as it developed into the Internet, there was no central control of what would happen and how it would happen. There is still no central governance that can force all hardware and software developers to act in a certain way. As a result, global decisions are complex, and require negotiations between a variety of stakeholders; and so they are rare. Most changes to the Internet are driven by the desire for new functionality, and those who want to make them must get buy-in from those who must also change. The size, distributed governance, and inertia of the Internet community, and the need for backwards compatibility mean that changes that have obvious benefits either cannot be made, or can only be made slowly.

Security is not the kind of property that can be added into a system that has already been designed. While the design of a bridge can be altered to make it bear a 30% greater load, a complex software/hardware system cannot really be altered to make it 30% more secure. The more-secure design must be started from the beginning, an impossibility when the system already exists. All the can be done is hardening, adding *ad hoc* features to make breaching security a little more difficult. There are perennial proposals to design and build Internet 2.0 but all have foundered on the sunk costs of the current Internet.

Much of the security in the Internet is built around the metaphor of a castle—high, protective walls, with gates the constrain data movement, allowing it only with authentication. Authentication continues to be a complex problem, because of the large number of identities in play, and because the point of authentication is typically far from the system being authenticated. Many kinds of data flow freely because there isn't a way to authenticate them robustly (email) or there are commercial reasons not to (the World Wide Web).

There is also little demand from Internet users for greater security, at least when counter-balanced by the lure of new Internet-based services. It is a truism that convenience always trumps security in practice.

The Internet began with an infrastructure that didn't pay much attention to cybersecurity; its explosive growth in scale and functionality provided little incentive or opportunity to add in security features; and its distributed nature and governance meant that there was no leverage to force better security in the face of developers and users who preferred other features. The difficulties of cybersecurity almost all trace back to this history and these drivers.

Cybersecurity that could (at least in theory) have been reduced to a tractable problem has become a war of attrition. The adversaries in this war are of three kinds: vandals, who are destructive for the fun of it; criminals, who use cyberspace as a platform and mechanism for making money from crime; and nations who use cyberspace as an intelligence-gathering platform and, increasingly, as a tool for national policy.

They are able to achieve their gains because of the weaknesses in the design and operation of cyberspace, but also because ordinary users of cyberspace are not sufficiently aware of the dangers. Much of what happens in cyberspace is invisible to ordinary users. This is, from one perspective, a success story since it hides great complexity from ordinary people; but it also means that they don't know which things they do in cyberspace are safe, which are risky, and which are dangerous. Education can help with this problem, but cyberspace is changing fast, so advice that was sensible a few years ago may not be today.

The fight to make cyberspace more secure has two prongs. The first is to continue to harden the hardware and software that instantiates cyberspace as much as possible. The problem is that hardening cyberspace is a global good, but inevitably costs some organizations much more than others—so it is underresourced, and good ideas fail for lack of access to the points where they could make a difference.

The second prong is to use law enforcement to penalize those who misuse cyberspace. There are many issues with this: the varying legal frameworks in different jurisdictions and the lag between the law and the exploits the criminals use; the skill sets of law enforcement personnel; and the ability of criminals to live in one country and commit crimes in another.

However, the biggest limitation to the law enforcement approach is the problem of associating online identities with real-world identities. The mechanisms for authentication online are not strong enough to tie online actions tightly to those who carried them out, certainly not to the level of "beyond a reasonable doubt". Those who carry out malicious acts in cyberspace have little need to fear criminal conviction.

Cybersecurity can be improved but by small, positive changes not by elegant sweeping large-scale actions.

Index